STUDENT WORKBOOK FOR
ELECTRONIC PRINCIPLES

FIFTH EDITION

▼

GAETANO A. GIUDICE
Queensborough Community College
of the City University of New York

GLENCOE
Macmillan/McGraw-Hill

New York, New York Columbus, Ohio Mission Hills, California Peoria, Illinois

Cover photographs:
 Special Effects, Charly Franklin/FPG International
 Microchip, Michael Simpson/FPG International
 Handmade Rice Paper, COMSTOCK, Inc./Michael Stuckey

Student Workbook for Electronic Principles, Fifth Edition

Copyright © 1993, 1991 by the Glencoe Division of Macmillan/McGraw-Hill School Publishing Company. All rights reserved. Except as permitted under the United States Copyright Act, no part of this publication may be reproduced or distributed in any form or by any means, or stored in a database or retrieval system, without the prior written permission of the publisher.

Send all inquiries to:
GLENCOE DIVISION
Macmillan/McGraw-Hill
936 Eastwind Drive
Westerville, OH 43081

ISBN 0-02-800850-2

Printed in the United States of America.

1 2 3 4 5 6 7 8 9   POH   01 00 99 98 97 96 95 94 93 92

CONTENTS

PREFACE		iv
Chapter 1	INTRODUCTION	1
Chapter 2	SEMICONDUCTORS	11
Chapter 3	DIODE THEORY	15
Chapter 4	DIODE CIRCUITS	27
Chapter 5	SPECIAL-PURPOSE DIODES	39
Chapter 6	BIPOLAR TRANSISTORS	47
Chapter 7	TRANSISTOR FUNDAMENTALS	55
Chapter 8	TRANSISTOR BIASING	67
Chapter 9	AC MODELS	81
Chapter 10	VOLTAGE AMPLIFIERS	91
Chapter 11	POWER AMPLIFIERS	103
Chapter 12	EMITTER FOLLOWERS	111
Chapter 13	FIELD EFFECT TRANSISTORS	121
Chapter 14	FET CIRCUITS	131
Chapter 15	THYRISTORS	145
Chapter 16	FREQUENCY EFFECTS	155
Chapter 17	OP-AMP THEORY	167
Chapter 18	MORE OP-AMP THEORY	175
Chapter 19	OP-AMP NEGATIVE FEEDBACK	181
Chapter 20	LINEAR OP-AMP CIRCUITS	187
Chapter 21	NONLINEAR OP-AMP CIRCUITS	195
Chapter 22	OSCILLATORS	197
Chapter 23	REGULATED POWER SUPPLIES	199
Chapter 24	COMMUNICATIONS SYSTEMS	201
Answers		203

PREFACE

This workbook was written to supplement the textbook *Electronic Principles*, Fifth Edition. Its purpose is to increase and strengthen student understanding of the material in the text. Each chapter of the workbook follows key sections of corresponding chapters in the text. It is recommended that the material in the text be studied before the student attempts to do the correlated exercises in the workbook.

Each chapter in the workbook contains true/false and completion questions. These can be used by students both to measure their progress and to reinforce their knowledge of the material in the text. Each question is followed by a reference number that indicates the section of the text covered by that question.

Most chapters in the workbook include both illustrative and practice problems. The illustrative problems are formatted into two columns. The column on the left contains a step-by-step procedure detailing what is required to solve the problem. On the right there are statements giving a rationale for the procedures used. Each step is presented in a clear and concise way, with the student being carefully guided through the problem-solving process. This helps students to formulate a problem-solving strategy. These illustrative problems may be used as models to help solve the practice problems. Answers to the practice problems are provided for students to use as a self-check. Some chapters also include sections on practical techniques, circuit highlights, or both. The sections on practical techniques provide some knowledge of measuring techniques. The sections on circuit highlights summarize key circuit concepts.

For further reference, most chapters include a number of additional problems. These problems are designed and sequenced to give students the practice they need to improve their problem-solving skills. Some problems show how a circuit behaves by changing the circuit parameters one at a time, thus enabling the students to see how the changes affect their answers. Understanding how a circuit behaves is of paramount importance for the troubleshooter. The problems are referenced to the corresponding section of the textbook chapters.

Answers to the true/false and completion questions and to the problems are provided in the back of the workbook. Answers to the problems may differ in some cases from those of the students depending on how the values are rounded.

It is hoped that this workbook will prove useful to students studying *Electronic Principles*. I welcome all comments and suggestions. Please address them to my attention: Electrical and Computer Engineering Technology Department, Queensborough Community College, Bayside, New York 11364.

I am very grateful to Dr. Albert P. Malvino, who made it possible for me to write this book, and to Brian Mackin and the staff at Glencoe for their diligent efforts to produce the best possible workbook. I am deeply indebted to my colleagues, Gabriel Koursourou, Bernard Mohr, and Henry Zanger. Their willingness to share their knowledge and skills has allowed me to grow as a technologist and as an educator. I would like to thank Mr. Ury Krotinsky and the entire staff of the Instructional Resources Center at Queensborough Community College for the use of their facilities. I would also like to thank Ulrich Zeisler and Frank Gergelyi, who reviewed the manuscript and contributed several valuable suggestions. Finally, for their encouragement and support in getting this workbook completed, I extend my sincere appreciation to my friends and family, and especially to Anita Scorcia.

Gaetano Antonio Giudice

CHAPTER 1
INTRODUCTION

Study Chap. 1 in *Electronic Principles*.

I. TRUE / FALSE

Answer true (T) or false (F) to each of the following statements.

Answer

1. An ideal voltage source has zero internal resistance. **(1-1)** 1. ___
2. A real voltage source has some internal resistance. **(1-1)** 2. ___
3. A stiff voltage source has internal resistance that is at least 100 times smaller than the load resistor. **(1-1)** 3. ___
4. An ideal current source has no internal resistance. **(1-2)** 4. ___
5. A real current source has a large internal resistance. **(1-2)** 5. ___
6. A stiff current source has an internal resistance that is at least 100 times smaller than the load resistor. **(1-2)** 6. ___
7. The Thevenin voltage equals the load voltage when the load resistor is disconnected. **(1-3)** 7. ___
8. The Thevenin resistance is dependent on the load resistance. **(1-3)** 8. ___
9. Thevenin proved mathematically that a complicated circuit facing a load can be replaced by an ideal voltage source and series resistor. **(1-3)** 9. ___
10. The Norton current equals the load current if the load resistance is zero. **(1-4)** 10. ___
11. The Norton current is independent of the load resistor. **(1-4)** 11. ___
12. Norton's theorem states that a complicated circuit facing a load can be replaced by an ideal current source and a series resistor. **(1-4)** 12. ___
13. A solder bridge may cause a short. **(1-5)** 13. ___
14. A cold-solder joint may cause a short. **(1-5)** 14. ___
15. An open resistor has zero current through it. **(1-5)** 15. ___
16. A shorted resistor has zero voltage across it. **(1-5)** 16. ___
17. The ideal approximation of a device is the simplest equivalent circuit of the device. **(1-6)** 17. ___
18. The third approximation of a device yields the least precision. **(1-6)** 18. ___
19. When troubleshooting, you will find the ideal approximation adequate. **(1-6)** 19. ___

Copyright © 1993 by the Glencoe Division of Macmillan/McGraw-Hill School Publishing Company. All rights reserved.

II. COMPLETION

Complete each of the following.

1. An ideal voltage source produces an output voltage that is ____. (1-1)
2. You can ignore the internal resistance of a ____ voltage source. (1-1)
3. A real voltage source has some ____ resistance. (1-1)
4. An ideal voltage source can exist only as a ____ device. (1-1)
5. An ideal current source has an internal resistance that is ____. (1-2)
6. The internal resistance of a real current source is in parallel with an ____ current source. (1-2)
7. A real current source works best when it has a very ____ internal resistance. (1-2)
8. The Thevenin voltage is sometimes called ____-____ voltage. (1-3)
9. In measuring the Thevenin resistance of a circuit, you must physically replace all voltage sources with ____ circuits. (1-3)
10. In measuring the Thevenin resistance of a circuit, you must physically replace all current sources with ____ circuits. (1-3)
11. The Norton current is sometimes called the ____-____ current. (1-4)
12. The Norton current can also be obtained by dividing the Thevenin voltage by the ____ resistance. (1-4)
13. The Norton resistance is equal to the ____ resistance. (1-4)
14. An open device is characterized by having ____ resistance. (1-5)
15. A shorted device is characterized by having ____ resistance. (1-5)
16. No current can flow through an ____ device. (1-5)
17. No voltage can exist across a ____ device. (1-5)
18. The simplest equivalent circuit of a device is the ____ approximation of the device. (1-6)
19. When the greatest precision is required from approximating, the ____ approximation method must be used. (1-6)
20. The ideal approximation of a connecting wire is a conductor of ____ resistance. (1-6)
21. When only some extra features are included to improve the analysis of an ideal approximation, the ____ approximation method is to be used. (1-6)

Answer

1. ____
2. ____
3. ____
4. ____
5. ____
6. ____
7. ____
8. ____
9. ____
10. ____
11. ____
12. ____
13. ____
14. ____
15. ____
16. ____
17. ____
18. ____
19. ____
20. ____
21. ____

III. ILLUSTRATIVE AND PRACTICE PROBLEMS

Illustrative Problem 1. Find the Thevenin equivalent circuit external to the load resistor for the given circuit.

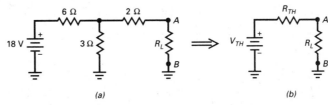

Figure 1-1. (a) Given circuit. (b) Thevenin equivalent circuit.

Steps	Comments
1. Find V_{TH}. **Figure 1-2** $V_{TH} = \dfrac{3\,\Omega}{6\,\Omega + 3\,\Omega}\, 18\text{ V} = 6\text{ V}$ voltage divider rule	**a.** Remove the load. **b.** Calculate the open-circuit voltage V_{TH}. **c.** The value of V_{TH} is the same as the voltage that appears across the 3-Ω resistor.
2. Find R_{TH}. **Figure 1-3** $R_{TH} = (3\,\Omega // 6\,\Omega) + 2\,\Omega = 4\,\Omega$	**a.** Replace the voltage source with a short. **b.** Calculate the equivalent resistance R_{TH} between terminals A and B with the load removed.
3. Draw the Thevenin equivalent circuit. **Figure 1-4**	Include the values of R_{TH} and V_{TH} on the drawing.

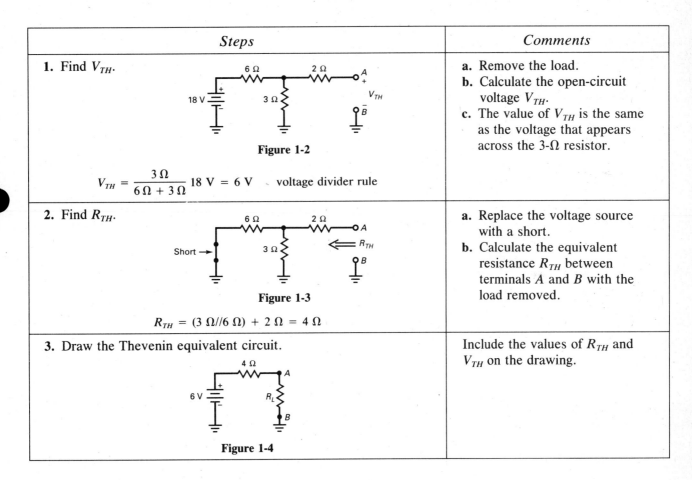

Practice Problem 1. Find the Thevenin equivalent circuit external to the load resistor for the following circuit.

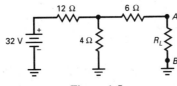

Figure 1-5

Answers: $V_{TH} = 8$ V, $R_{TH} = 9\,\Omega$.

Illustrative Problem 2. Find the Norton equivalent circuit external to the load resistor for the given circuit.

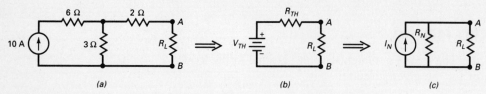

Figure 1-6. (a) Given circuit. (b) Thevenin equivalent circuit. (c) Norton equivalent circuit.

Steps	Comments
1. Find V_{TH}. *Figure 1-7* $V_{TH} = (10 \text{ A})(3 \text{ }\Omega) = 30 \text{ V}$ Ohm's law	First find the Thevenin equivalent circuit: a. Remove the load. b. Calculate the open-circuit voltage V_{TH}. c. The value of V_{TH} is the same as the voltage that appears across the 3-Ω resistor.
2. Find R_{TH}. *Figure 1-8* $R_{TH} = 3 \text{ }\Omega + 2 \text{ }\Omega = 5 \text{ }\Omega$	c. Replace the current source with an open circuit. d. Calculate the equivalent resistance R_{TH} between terminals A and B with the load removed.
3. Find I_N. $I_N = \dfrac{V_{TH}}{R_{TH}} = \dfrac{30 \text{ V}}{5 \text{ }\Omega} = 6 \text{ A}$	Convert the Thevenin circuit to a Norton circuit.
4. Find R_N. $R_N = R_{TH} = 5 \text{ }\Omega$	
5. Draw the Norton equivalent circuit. *Figure 1-9*	Include the values of I_N and R_N on the drawing.

Practice Problem 2. Find the Norton equivalent circuit external to the load resistor for the given circuit.

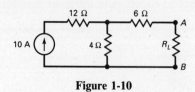

Figure 1-10

Answers: $I_N = 4 \text{ A}$, $R_N = 10 \text{ }\Omega$.

IV. PRACTICAL TECHNIQUES

Find the Thevenin equivalent circuit between terminals A and B using voltage measurements.

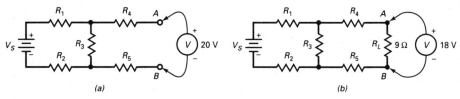

Figure 1-11. (a) No load voltage measurements. (b) Full load voltage measurements.

Steps	Comments
1. Find V_{TH}. $V_{TH} = 20$ V	V_{TH} is equal to the open circuit voltage measurement.
2. Find R_{TH}. **Figure 1-12** $I_L = \dfrac{18 \text{ V}}{9 \text{ }\Omega} = 2$ A Voltage drop across $R_{TH} = 20$ V $-$ 18 V $=$ 2 V $R_{TH} = \dfrac{2 \text{ V}}{2 \text{ A}} = 1$ Ω	a. Draw the Thevenin equivalent circuit. b. Calculate the current through the load (I_L). c. Calculate the voltage drop across the Thevenin resistance. d. Use Ohm's law to calculate R_{TH}.

Practice Problem. Find the Thevenin equivalent circuit between terminals A and B using voltage measurements.

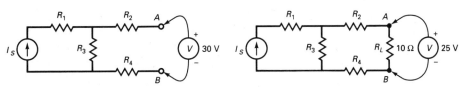

Figure 1-13

Answers: $V_{TH} = 30$ V, $R_{TH} = 2$ Ω.

V. PROBLEMS

Sec. 1-1 Voltage Sources

1-1. For the circuit of Fig. 1-14:
  a. What is the reading of the voltmeter when the load resistor R_L is removed?
  b. Determine the readings of the voltmeter when the load resistor R_L is set to each of the following values: 1 Ω, 10 Ω, 100 Ω, 1000 Ω, and 10,000 Ω.
  c. With respect to part b, which values of R_L make the voltage source appear stiff?
  d. With respect to parts b and c, does the stiff voltage source condition produce voltmeter readings that are closer in value to the ideal voltage source?

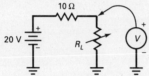

Figure 1-14.

1-2. A 9-V battery has an internal resistance of 0.1 Ω. For what values of load resistance does the battery voltage appear stiff?

1-3. In the circuit of Fig. 1-15, the reading of the voltmeter is 11.5 V.
  a. Find the internal resistance R_S of the voltage source.
  b. Does the voltage source appear stiff?

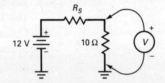

Figure 1-15.

1-4. A 1.5-V battery has an internal resistance of 0.5 Ω and draws 20 mA when a light bulb is connected across it.
  a. Find the voltage drop across the battery's internal resistance.
  b. Find the voltage drop across the bulb.
  c. Under these operating conditions, does the battery voltage appear to be stiff?

1-5. A 12-V car battery has an internal resistance of 0.05 Ω and draws 50 A when it is used to start a car.
  a. What is the load resistance R_L of the car's starter?
  b. Under these conditions, does the battery voltage appear to be stiff?

Sec. 1-2 Current Sources

1-6. For the circuit of Fig. 1-16:
  a. What is the reading of the milliammeter when the load resistor R_L is shorted?
  b. What are the readings of the milliammeter when the load resistor R_L is set to each of the following values: 5 Ω, 50 Ω, 500 Ω, 5 kΩ, and 50 kΩ?
  c. With respect to part *b*, which values of R_L make the current source appear stiff?
  d. With respect to parts *b* and *c*, does the stiff current source condition produce milliammeter readings that are closer in value to the ideal current source?

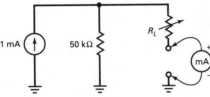

Figure 1-16.

1-7. In the circuit of Fig. 1-17, the reading of the milliammeter is 19 mA.
  a. Find the internal resistance R_S of the current source.
  b. Does the current source appear stiff?

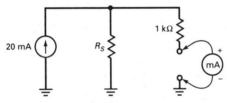

Figure 1-17.

Sec. 1-3 Thevenin's Theorem

1-8. In the circuit of Fig. 1-18, the voltmeter reading of the open-circuit voltage is 10 V and the ammeter reading of the short-load current is 5 A.
  a. Find the Thevenin voltage.
  b. Find the Thevenin resistance.

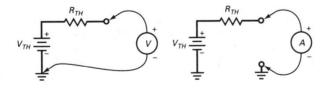

Figure 1-18.

*1-9. In the circuit of Fig. 1-19, the voltmeter of the open-circuit voltage (R_L removed) is 20 V. The voltmeter reading is 18 V, when R_L is 10 Ω.
  a. Find the Thevenin voltage.
  b. Find the Thevenin resistance.

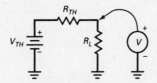

Figure 1-19.

1-10. With respect to terminals A and B in the circuit of Fig. 1-20, do the following:
  a. Find the Thevenin voltage.
  b. Find the Thevenin resistance.
  c. Draw the Thevenin equivalent circuit.

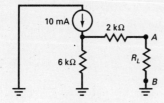

Figure 1-20.

*1-11. In the circuit of Fig. 1-21, the voltmeter reading of the open-circuit voltage (R_L removed) is 8 V. The voltmeter reading is 5 V, when R_L is 100 Ω.
  a. Find the Thevenin voltage.
  b. Find the Thevenin resistance.
  c. Draw the Thevenin equivalent circuit.

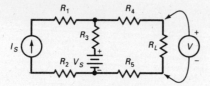

Figure 1-21.

*1-12. In the circuit of Fig. 1-22, the voltmeter reading of the open-circuit voltage is 5 V. The voltmeter reading is 4.8 V when a 10-Ω resistor is put across the dc power supply.
  a. Find the Thevenin voltage.
  b. Find the Thevenin resistance.
  c. Draw the Thevenin equivalent circuit of the dc power supply.

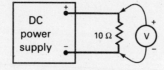

Figure 1-22.

* See "Practical Techniques."

Sec. 1-4 Norton's Theorem

1-13. Using the circuit and meter measurements that are given in Prob. 1-8:
  a. Find the Norton current.
  b. Find the Norton resistance.
  c. Draw the Norton equivalent circuit.

1-14. Using the circuit and meter measurements that are given in Prob. 1-9:
  a. Find the Norton current.
  b. Find the Norton resistance.
  c. Draw the Norton equivalent circuit.

1-15. With respect to terminals A and B in the circuit of Fig. 1-20, do the following:
  a. Find the Norton current.
  b. Find the Norton resistance.
  c. Draw the Norton equivalent circuit.

1-16. Using the circuit and meter measurements that are given in Prob. 1-11:
  a. Find the Norton current.
  b. Find the Norton resistance.
  c. Draw the Norton equivalent circuit.

1-17. With respect to terminals A and B in the circuit of Fig. 1-23, do the following:
  a. Find the Norton current.
  b. Find the Norton resistance.
  c. Draw the Norton equivalent circuit.

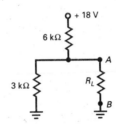

Figure 1-23.

*1-18. In the circuit of Fig. 1-24, the voltmeter of the open-circuit voltage is 30 V. The voltmeter reading is 25 V, when R_L is 1 kΩ.
  a. Find the Norton current.
  b. Find the Norton resistance.
  c. Draw the Norton equivalent circuit.

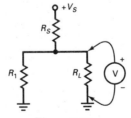

Figure 1-24.

* See "Practical Techniques."

Sec. 1-5 Troubleshooting

1-19. For the circuit of Fig. 1-25, determine the readings of the voltmeters for the circuit conditions that are given in Table 1-1. Fill in Table 1-1 with your answers.

Trouble	Voltmeter Readings		
	V_A	V_B	V_C
Circuit OK			
R_1 shorted			
R_1 opened			
R_2 shorted			
R_2 opened			
R_3 shorted			
R_3 opened			
Power supply off			

Table 1-1

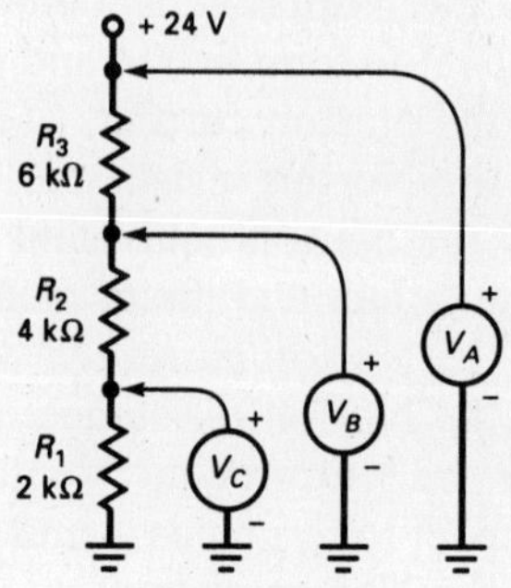

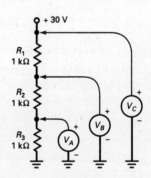

Figure 1-25

1-20. For the circuit of Fig. 1-26, determine the circuit's probable troubles from the voltmeter's readings given in Table 1-2. Fill in Table 1-2 with your answers.

Probable Trouble	Voltmeter Readings		
	V_A, V	V_B, V	V_C, V
	0	30	30
	15	30	30
	10	20	30
	0	0	30
	0	15	30
	0	0	0

Table 1-2

Figure 1-26

CHAPTER 2

SEMICONDUCTORS

Study Chap. 2 in *Electronic Principles*.

I. TRUE / FALSE

Answer true (T) or false (F) to each of the following statements.   *Answer*

1. Valence electrons that are liberated from their atoms are called free electrons. **(2-1)**   1. ___
2. Elements whose atoms have eight valence electrons are the best conductors. **(2-1)**   2. ___
3. Electrically charged atoms are called ions. **(2-1)**   3. ___
4. The most widely used semiconductor material is germanium. **(2-2)**   4. ___
5. Germanium has the same number of valence electrons as silicon. **(2-2)**   5. ___
6. The chemical bond that causes silicon atoms to combine and form a crystal is known as an ionic bond. **(2-3)**   6. ___
7. A hole is created in a pure silicon crystal when an electron breaks away from a covalent bond. **(2-3)**   7. ___
8. At room temperature, a pure silicon crystal acts almost like a perfect insulator. **(2-3)**   8. ___
9. If the temperature of a silicon crystal is increased, the number of free electrons and holes will decrease. **(2-3)**   9. ___
10. A pure silicon crystal always has the same number of free electrons and holes. **(2-3)**   10. ___
11. An intrinsic semiconductor contains impurities. **(2-4)**   11. ___
12. A semiconductor can only support one distinct type of flow. **(2-4)**   12. ___
13. Free electrons and holes move in the same direction. **(2-5)**   13. ___
14. An extrinsic semiconductor contains impurities. **(2-6)**   14. ___
15. Doping is a process of removing impurities from a semiconductor. **(2-6)**   15. ___
16. A lightly doped semiconductor has less resistance than a heavily doped semiconductor. **(2-6)**   16. ___
17. Pentavalent atoms are added to pure molten silicon to increase the silicon crystal's number of free electrons. **(2-6)**   17. ___
18. Trivalent atoms are added to pure molten silicon to decrease the silicon crystal's number of holes. **(2-6)**   18. ___
19. Because of its advantages, germanium is the most popular and useful semiconductor material. **(2-6)**   19. ___

20. Silicon that has been doped with a trivalent impurity is called an *n*-type semiconductor. (2-7)   20. ___
21. There are more free electrons than holes in an *n*-type semiconductor. (2-7)   21. ___
22. The majority carriers in a *p*-type semiconductor are holes. (2-7)   22. ___
23. The minority carriers in an *n*-type semiconductor are free electrons. (2-7)   23. ___
24. At room temperature, the barrier potential for silicon diodes is approximately 0.3 V. (2-8)   24. ___
25. The barrier potential exists across the depletion layer of a *pn* crystal. (2-8)   25. ___
26. Ions in the depletion layer produce the barrier potential. (2-8)   26. ___
27. Current flows easily in a forward-biased diode. (2-9)   27. ___
28. When an external voltage aids the barrier potential, the diode is forward-biased. (2-9)   28. ___
29. The amount of forward-bias voltage applied to a diode before it can conduct must be greater than the barrier potential. (2-9)   29. ___
30. A reverse-biased diode acts approximately like an open switch. (2-10)   30. ___
31. When an external voltage aids the barrier potential, the diode is reverse-biased. (2-10)   31. ___
32. The width of the depletion layer decreases when a diode is reverse-biased. (2-10)   32. ___
33. A germanium diode has a much larger saturation current than a silicon diode of the same size and shape. (2-10)   33. ___
34. A diode conducts heavily when the applied reverse voltage exceeds the diode's breakdown voltage. (2-11)   34. ___
35. Most diodes are not designed to operate in their breakdown region. (2-11)   35. ___
36. The breakdown voltage of a diode depends on how heavily the diode is doped. (2-11)   36. ___
37. The value of the barrier potential depends on the junction temperature. (2-14)   37. ___
38. An increase in junction temperature will cause the width of the depletion to increase. (2-14)   38. ___
39. The saturation current is not dependent on the junction temperature. (2-15)   39. ___
40. The surface-leakage current is dependent on the reverse voltage. (2-15)   40. ___

II. COMPLETION

Complete each of the following.

Answer

1. In a copper atom, the strength of the force of attraction acting on the valence electron by the nucleus is ____. (2-1)   1. _____
2. The outer orbit of an atom is called a ____ orbit. (2-1)   2. _____
3. If a neutral atom gains electrons, it becomes negatively charged and is called a negative ____. (2-1)   3. _____
4. A semiconductor is an element with a valence of ____. (2-2)   4. _____
5. An element's electrical conductivity is determined by the number of electrons in its ____ orbit. (2-2)   5. _____
6. When silicon atoms combine to form a crystal, the number of electrons each atom shares in its valence orbit is ____. (2-3)   6. _____
7. Each atom in a silicon crystal shares its electrons with four neighboring atoms to form ____ bonds. (2-3)   7. _____
8. The merging of a free electron and a hole in a silicon crystal is called ____. (2-3)   8. _____
9. The amount of time between the creation and disappearance of a free electron and a hole is called the ____. (2-3)   9. _____

10. Some free electrons and holes are being created in a silicon crystal by _____ energy. (2-3)
11. If a semiconductor is only made up of like atoms, it is called an _____ semiconductor. (2-4)
12. The flow of holes is created by the movement of _____ electrons. (2-4)
13. Because free electrons and holes can carry a charge from one place to another, they are often called _____. (2-5)
14. The process of adding impurities to an intrinsic semiconductor to increase its conductivity is called _____. (2-6)
15. Each trivalent atom is called an _____ atom, because it contributes a hole in a silicon crystal that can accept an electron. (2-6)
16. Each pentavalent atom is called a _____ atom, because it donates an extra electron in the silicon crystal. (2-6)
17. Silicon that has been doped with a pentavalent impurity is called a _____-type semiconductor. (2-7)
18. The number of holes is greater than the number of free electrons in a _____-type semiconductor. (2-7)
19. In an *n*-type semiconductor, the free electrons are called _____ carriers. (2-7)
20. In a *p*-type semiconductor, the holes are called _____ carriers. (2-7)
21. In an *n*-type semiconductor, the holes are called _____ carriers. (2-7)
22. Another name for the *pn* crystal is the _____ diode. (2-8)
23. The region near the *pn* junction that is emptied of carriers is called the _____ layer. (2-8)
24. At room temperature, the barrier potential is approximately 0.7 V for _____ diodes. (2-8)
25. In order for a germanium diode to conduct when it is forward-biased, the applied voltage must be greater than _____ V. (2-9)
26. In order for a silicon diode to conduct when it is forward-biased, the applied voltage must be greater than _____ V. (2-9)
27. When the reverse voltage increases, the width of the depletion layer _____. (2-10)
28. The amount of current flowing through a diode that is reverse-biased is approximately _____. (2-10)
29. The reverse current caused by the thermally produced minority carriers is called the _____ current. (2-10)
30. The reverse current caused by surface impurities and imperfections in the crystal is called _____-_____ current. (2-10)
31. Diode breakdown caused by high-speed minorities carriers knocking valence electrons loose is known as the _____ effect. (2-11)
32. Diode breakdown caused by a high electric field intensity that is created across the depletion layer and pulls electrons out of their valence orbit is known as the _____ effect. (2-11)
33. It can be estimated that for either a germanium or a silicon diode, the barrier potential decreases _____ for each 1°C rise in temperature. (2-14)
34. The saturation current doubles for every _____ °C rise in temperature. (2-15)

10. _____
11. _____
12. _____
13. _____
14. _____
15. _____
16. _____
17. _____
18. _____
19. _____
20. _____
21. _____
22. _____
23. _____
24. _____
25. _____
26. _____
27. _____
28. _____
29. _____
30. _____
31. _____
32. _____
33. _____
34. _____

III. PROBLEMS

Sec. 2-14 Barrier Potential and Temperature

2-1. Complete Table 2-1 for the conditions that are given on a silicon and germanium diode.

Junction Temperature, °C	Silicon Barrier Potential, V	Germanium Barrier Potential, V
5		
15		
25	0.7	0.3
50		
75		

Table 2-1

Sec. 2-15 Reverse-Biased Diode

2-2. Complete Table 2-2 for the conditions that are given on a silicon and germanium diode.

Junction Temperature, °C	Silicon Saturaton Current, nA	Germanium Saturation Current, µA
5		
15		
25	4	2
35		
55		

Table 2-2

2-3. Complete Table 2-3 for the conditions that are given on a silicon and germanium diode.

Reverse Voltage across the Diode, V	Silicon Surface-Leakage Current, nA	Germanium Surface-Leakage Current, nA
1		
5		
10	10	50
20		
40		

Table 2-3

Name _____ Date _____

CHAPTER 3
DIODE THEORY

Study Chap. 3 in *Electronic Principles*.

I. TRUE / FALSE

Answer true (T) or false (F) to each of the following statements.

Answer

1. The diode symbol looks like an arrow that points from the *n* side to the *p* side of the diode. (3-1)
1. ___

2. A resistor is a linear device because a plot of current versus voltage is a straight line. (3-1)
2. ___

3. If circuit conditions force conventional current to flow in the same direction as the diode arrow, then the diode is forward-biased. (3-2)
3. ___

4. The bulk resistance of a diode depends on the doping level and the size of the *p* and *n* regions. (3-3)
4. ___

5. On the graph of the diode curve, the voltage where the current starts to increase rapidly is called the knee voltage. (3-3)
5. ___

6. A current-limiting resistor is used to keep the diode current less than its maximum value. (3-3)
6. ___

7. The product of the diode voltage and current equals to the power dissipated by the diode. (3-3)
7. ___

8. When a diode is reverse-biased, it conducts a small leakage current. (3-4)
8. ___

9. The ideal diode is the first approximation of the diode. (3-5)
9. ___

10. An ideal diode that is forward-biased acts like a closed switch. (3-5)
10. ___

11. An ideal diode dissipates power. (3-5)
11. ___

12. For the second approximation, when a silicon diode is turned on, it can be replaced by a barrier potential of 0.7 V. (3-6)
12. ___

13. A silicon diode can be replaced by an open circuit when the source voltage is less than 0.7 V or when the diode is reverse-biased. (3-6)
13. ___

14. For the third approximation, when a silicon diode is turned on, it can be replaced by a barrier potential of 0.7 V in parallel with its bulk resistance. (3-7)
14. ___

15. The bulk resistance of a diode is usually large. (3-7)
15. ___

16. The second approximation is an excellent compromise to use whenever you are in doubt about which approximation to use. (3-8)
16. ___

17. The first approximation method is often used in troubleshooting. (3-8)
17. ___

Copyright © 1993 by the Glencoe Division of Macmillan/McGraw-Hill School Publishing Company. All rights reserved.

18. When using an ohmmeter to check a silicon diode, the reverse resistance should be greater than 1000 times the forward resistance. (3-9)   18. ___
19. Practicing up-down thinking helps develop a better understanding of circuit behavior. (3-10)   19. ___
20. Up-down thinking has very little to do with how circuit variables are related. (3-10)   20. ___
21. The maximum ratings that are provided on a manufacturer's data sheet can be exceeded without the danger of destroying the semiconductor component. (3-11)   21. ___

II. COMPLETION

Complete each of the following.

Answer

1. The p side of a diode is called the ___. (3-1)   1. ___
2. The n side of a diode is called the ___. (3-1)   2. ___
3. If the external circuit is trying to push free electrons in the direction that is opposite the diode's arrow, then the diode is ___-biased. (3-2)   3. ___
4. Because a diode's current is not directly proportional to its voltage, it is called a ___ device. (3-3)   4. ___
5. The maximum forward current is one of the maximum ratings that is given on the manufacturer's ___ sheet. (3-3)   5. ___
6. The resistance of the p and n region that causes large current changes for small voltages is called the ___ resistance of the diode. (3-3)   6. ___
7. Once the diode has fired, the diode's barrier potential is equal to the ___ voltage shown on the diode curve. (3-3)   7. ___
8. The diode current is very small for all reverse voltages less than the diode's ___ voltage. (3-4)   8. ___
9. When an ideal diode is reverse-biased, it acts like an ___ switch. (3-5)   9. ___
10. The equivalent circuit for the second approximation of a silicon diode that is forward-biased is a closed switch in series with a barrier potential of ___. (3-6)   10. ___
11. The phrase "the diode turns on" means that the diode is properly forward-biased so that it can ___. (3-6)   11. ___
12. The diode curve is affected by the diode's ___ resistance. (3-7)   12. ___
13. To minimize error, the third approximation should be used when the size of the load resistance in series with the diode is ___. (3-8)   13. ___
14. The largest errors will be encountered when one is using the ___ approximation. (3-8)   14. ___
15. Ohmmeter readings of a diode that show low resistance in both directions indicate that the diode is ___. (3-9)   15. ___
16. Ohmmeter readings of a diode that show high resistance in both directions indicate that the diode is ___. (3-9)   16. ___
17. Up-down thinking deals with how changes in the circuit's independent variables affect changes in the circuit's ___ variables. (3-10)   17. ___
18. The rating of the dc blocking voltage is another way of indicating the diode reverse ___ voltage. (3-11)   18. ___
19. Failure studies of devices show that the lifetime of a device decreases the closer you get to the ___ rating. (3-11)   19. ___

III. ILLUSTRATIVE AND PRACTICE PROBLEMS

Illustrative Problem 1. For the silicon diode that is shown in Fig. 3-1, use all three approximation methods to find:
  a. The current through the diode.
  b. V_o (output voltage).
  c. The power dissipated by the diode.

Data: $r_B = 1 \, \Omega$

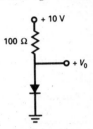

Figure 3-1

Steps	Comments
1. First approximation (ideal diode). Figure 3-2 $I = \dfrac{10 \text{ V}}{100 \, \Omega} = 100 \text{ mA}$ $V_o = 0 \text{ V} \quad \text{(short)}$ $P_D = (100 \text{ mA})(0 \text{ V}) = 0 \text{ W}$	a. Diode is forward-biased. (Circuit conditions are forcing conventional current to flow in the direction of the diode arrow.) b. Replace the diode with a short. c. Solve.
2. Second approximation. Figure 3-3 $I = \dfrac{10 \text{ V} - 0.7 \text{ V}}{100 \, \Omega} = 93 \text{ mA}$ $V_o = 0.7 \text{ V}$ $P_D = (93 \text{ mA})(0.7 \text{ V}) = 65.1 \text{ mW}$	a. Diode is forward-biased. b. Replace the diode with a barrier potential of 0.7 V with a polarity to oppose current flowing through it from the supply voltage. c. Solve.
3. Third approximation. Figure 3-4 $I = \dfrac{10 \text{ V} - 0.7 \text{ V}}{100 \, \Omega + 1 \, \Omega} = 92.1 \text{ mA}$ $V_o = 0.7 \text{ V} + (92.1 \text{ mA})(1 \, \Omega) = 0.792 \text{ V}$ $P_D = (92.1 \text{ mA})(0.792 \text{ V}) = 72.9 \text{ mW}$	a. Diode is forward-biased. b. Replace the diode with a barrier potential of 0.7 V in series with its bulk resistance r_B. The polarity of the barrier potential should be the same as step 2b. c. Solve.

Practice Problem 1. For the silicon diode that is shown in Fig. 3-5, use all three approximation methods to find:
a. The current through the diode.
b. V_o (output voltage).
c. The power dissipated by the diode.

Data: $r_B = 0.1\ \Omega$

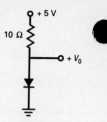

Figure 3-5

Answers:

	First Approximation	Second Approximation	Third Approximation
Current	500 mA	430 mA	426 mA
V_o	0 V	0.7 V	0.743 V
P_D	0 W	301 mW	317 mW

Illustrative Problem 2. For the silicon diode that is shown in Fig. 3-6, use all three approximation methods to find:
a. The current through the diode.
b. V_o (output voltage).
c. The power dissipated by the diode.

Data: $r_B = 1\ \Omega$

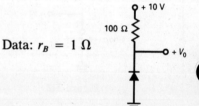

Figure 3-6

Steps	Comments
1. First approximation (ideal diode). $I = 0\ \text{A}$ open circuit $V_o = 10\ \text{V}$ open-circuit voltage $P_D = (0\ \text{A})(10\ \text{V}) = 0\ \text{W}$ Figure 3-7	a. Diode is reverse-biased. (Circuit conditions are forcing conventional current to flow in the opposite direction of the diode arrow.) b. Replace the diode with an open circuit. c. Solve.

18

Steps	Comments
2. Second approximation. *Figure 3-8* $I = 0\,A$ open circuit $V_o = 10\,V$ open-circuit voltage $P_D = (0\,A)(10\,V) = 0\,W$	**a.** Diode is reverse-biased. **b.** Replace the diode with an open circuit. **c.** Solve.
3. Third approximation. *Figure 3-9* $I = 0\,A$ open circuit $V_o = 10\,V$ open-circuit voltage $P_D = (0\,A)(10\,V) = 0\,W$	**a.** Diode is reverse-biased. **b.** Replace the diode with an open circuit. **c.** Solve.

Practice Problem 2. For the silicon diode that is shown in Fig. 3-10, use all three approximation methods to find:
  a. The current through the diode.
  b. V_o (output voltage).
  c. The power dissipated by the diode.

Data: $r_B = 0.1\,\Omega$

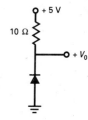

Figure 3-10

Answers:

	First Approximation	Second Approximation	Third Approximation
Current	0 A	0 A	0 A
V_o	5 V	5 V	5 V
P_D	0 W	0 W	0 W

Copyright © 1993 by the Glencoe Division of Macmillan/McGraw-Hill School Publishing Company. All rights reserved.

IV. PRACTICAL TECHNIQUES

1. Measure the leakage current I_R of a diode.

Steps	Comments
1. Construct the circuit that is shown in Fig. 3-11. **Figure 3-11**	a. The value of R should be at least 100 kΩ but not large enough to cause the voltmeter to load down the circuit. b. In order to insure that the voltage drop across R can be read easily, increase the voltage level of V_S. V_S cannot be set to a voltage level that exceeds the reverse breakdown voltage of the diode.
2. Measure the voltage across the resistor. Voltmeter reading = _____	This voltage measurement should be taken accurately.
3. Measure I_R. $I_R = \dfrac{\text{voltmeter reading}}{R}$	a. The current through the diode is equal to the current through the resistor. b. Using Ohm's Law, measure the current through the resistor.

Example 1. Using voltage measurements, find the leakage current I_R of the diode that is in Fig. 3-12.

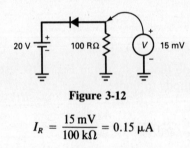

Figure 3-12

$$I_R = \frac{15 \text{ mV}}{100 \text{ k}\Omega} = 0.15 \text{ }\mu\text{A}$$

2. Identify an unmarked diode's anode and cathode.

Steps	Comments
1. Connect the analog ohmmeter across the diode as shown in Fig. 3-13 and take a reading. **Figure 3-13**	**a.** Identify the diode's leads. **b.** Use the ohmmeters $R \times 100$ scale. **c.** A low reading of about a few hundred ohms indicates the diode is forward-biased. Therefore, lead 1 is connected to the anode and lead 2 is connected to the cathode. **d.** A high reading of something greater than 1 MΩ indicates that the diode is reverse-biased. Therefore, lead 1 is connected to the cathode and lead 2 is connected to the anode.
2. Reverse the analog ohmmeter connections across the diode as shown in Fig. 3-14 and take another reading. **Figure 3-14**	Readings should change from low to high or vice versa.

Example. Determine what side of an unmarked diode is the anode and cathode.

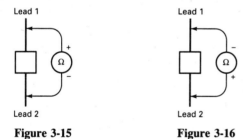

Figure 3-15   **Figure 3-16**

In Fig. 3-15, the ohmmeter reading is 600 Ω. In Fig. 3-16, the ohmmeter reading is ∞ Ω. Therefore, lead 1 is connected to the anode and lead 2 is connected to the cathode.

V. PROBLEMS

Sec. 3-3 The Forward Region

3-1. The voltmeter reading is 0.7 V in the circuit of Fig. 3-17. Find the current through the diode.

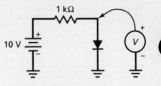

Figure 3-17

3-2. In the circuit of Fig. 3-18, the voltmeter reading is 0.68 V and the milliampere meter reading is 100 mA. How much power is the diode dissipating?

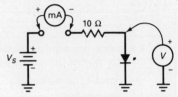

Figure 3-18

3-3. Find the value of R that would limit the current through the diode to 50 mA in the circuit of Fig. 3-19.

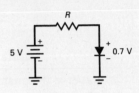

Figure 3-19

Sec. 3-5 The Ideal Diode

(Use the first approximation method for all diode circuits in this section.)

3-4. In the circuits of Fig. 3-20, determine whether the diode in each circuit is forward- or reverse-biased.

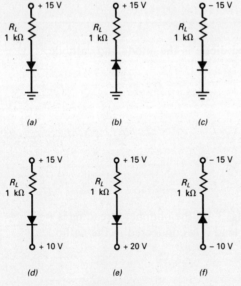

Figure 3-20

3.5 In the circuit of Fig. 3-20a, find:
   a. The diode current.
   b. The diode voltage.
   c. The load voltage.
   d. The diode power.
   e. The load power.
   f. The total power.

3-6. Repeat Prob. 3-5 for the circuit of Fig. 3-20b.

3-7. Repeat Prob. 3-5 for the circuit of Fig. 3-20d.

3-8. Repeat Prob. 3-5 for the circuit of Fig. 3-20e.

3-9. In the circuit of Fig. 3-21, determine:
  a. The current through the 10-kΩ resistor.
  b. The current through the 1-kΩ resistor.
  c. The current through the diode.
  d. The current through the 1-kΩ resistor with the diode connected in the reverse direction.

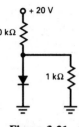

Figure 3-21

3-10. In the circuit of Fig. 3-22, determine:
  a. The current through the 1-kΩ resistor.
  b. The current through the diodes.

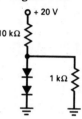

Figure 3-22

3-11. In the circuit of Fig. 3-23, find V_o.

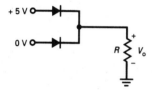

Figure 3-23

3-12. In the circuit of Fig. 3-24, find V_o.

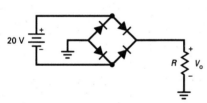

Figure 3-24

3-13. In the circuit of Fig. 3-24, reverse the polarity of the voltage source and find V_o.

3-14. Determine the reading of the voltmeter in the circuit of Fig. 3-25.

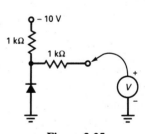

Figure 3-25

3-15. In the circuit of Fig. 3-26, find the voltages V_A and V_B for the following circuit conditions:
  a. $R = 0\ \Omega$ (shorted).
  b. $R = 5\ \text{k}\Omega$.
  c. $R = 26\ \text{k}\Omega$.
  d. $R = \infty\ \Omega$ (opened).

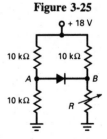

Figure 3-26

Sec. 3-6 The Second Approximation

(Use the second approximation method for all diode circuits in this section.)

3-16. The diode is silicon in the circuit of Fig. 3-20a. Find:
- a. The diode current.
- b. The diode voltage.
- c. The load voltage.
- d. The diode power.
- e. The load power.
- f. The total power.

3-17. Repeat Prob. 3-16 using a germanium diode.
3-18. Repeat Prob. 3-16 for the circuit of Fig. 3-20b.
3-19. Repeat Prob. 3-16 for the circuit of Fig. 3-20d.
3-20. Repeat Prob. 3-9 using a silicon diode.
3-21. Repeat Prob. 3-10 using silicon diodes.
3-22. Repeat Prob. 3-12 using a silicon diode.
3-23. Repeat Prob. 3-14 using a silicon diode.

Sec. 3-7 The Third Approximation

(Use the third approximation method for all diode circuits in this section.)

3-24. In the circuit of Fig. 3-27 (the diode is silicon):   Data: $r_B = 0.5\ \Omega$
- a. Find the diode current.
- b. Find the diode voltage.
- c. Find the diode power.

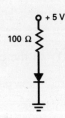

Figure 3-27

3-25. Repeat Prob. 3-24 with the polarity of the diode reversed.
3-26. The bulk resistance of each silicon diode is $0.25\ \Omega$ in the circuit of Fig. 3-28. Which diode will have a voltage drop across it closer to 0.7 V?

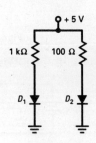

Figure 3-28

Sec. 3-9 Troubleshooting

3-27. Fill in the circuit's probable trouble in Table 3-1 for the given voltmeter readings of the circuit of Fig. 3-29.

Probable Trouble	Voltmeter Readings	
	V_1, V	V_2, V
	10	0
	10	5
	10	0.7

Table 3-1

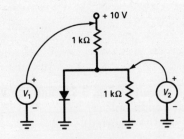

Figure 3-29

3-28. Fill in the diode's probable trouble on Table 3-2 for the given analog ohmmeter readings of the circuit of Fig. 3-30.

Probable Trouble	Ohmmeter Readings	
	Fig. 3-30a	Fig. 3-30b
	∞ Ω	∞ Ω
	100 Ω	100 Ω
	∞ Ω	500 Ω

Table 3-2

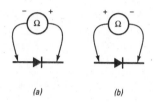

Figure 3-30

*3-29. In the circuit of Fig. 3-31, the reading of the analog ohmmeter is ∞ Ω. When the polarity of the analog ohmmeter is reversed, the reading is 540 Ω. Identify the anode and cathode of the unmarked diode.

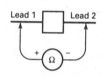

Figure 3-31

Sec. 3-10 Up-Down Thinking

3-30. Some of the independent variables are increased in the circuit of Fig. 3-32. As a result of this, indicate how the current through each element and the voltage at node A are affected. Complete Table 3-3 by inserting in the appropriate boxes a U for goes up, a D for goes down, and an N for shows no change. (Second approximation.)

	V_A	I_1	I_2	I_D
V_S increases				
R_1 increases				
R_2 increases				

Table 3-3

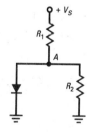

Figure 3-32

* See "Practical Techniques."

Sec. 3-15 Load Lines

3-31. For the circuit and diode characteristics of Fig. 3-33:
  a. Construct a load line for the circuit on the diode characteristics.
  b. Identify the Q-point.
  c. From the constructed load line, determine the values of the diode current and voltage.

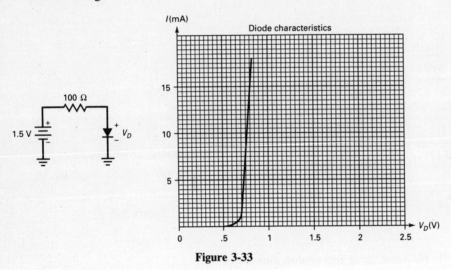

Figure 3-33

3-32. From the circuit's load line constructed on the diode characteristics in Fig. 3-34:
  a. Find the value of V_S.
  b. Find the value of R_S.
  c. Find the diode current.
  d. Find the diode voltage.

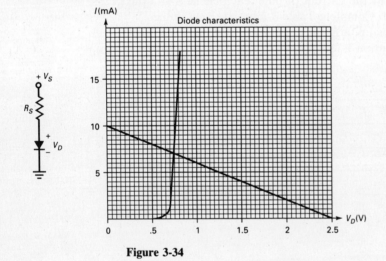

Figure 3-34

CHAPTER 4
DIODE CIRCUITS

Study Chap. 4 in *Electronic Principles*.

I. TRUE / FALSE

Answer true (T) or false (F) to each of the following statements.

Answer

1. In a step-up transformer the number of primary windings is greater than the number of secondary windings. **(4-1)** 1. ___
2. In an ideal transformer the coefficient of coupling is one and there are no power losses in its windings or core. **(4-1)** 2. ___
3. The load current of a half-wave rectifier flows for the entire cycle. **(4-2)** 3. ___
4. The frequency of the half-wave rectified signal is one-half the line frequency. **(4-2)** 4. ___
5. The ideal peak value of the load voltage of a full-wave rectifier is equal to the peak value of the secondary voltage. **(4-3)** 5. ___
6. The dc value of a full-wave rectified signal is 0.318 of its ideal peak value. **(4-3)** 6. ___
7. The ideal diode or the second approximation method is adequate in analyzing most full-wave circuits. **(4-3)** 7. ___
8. The ideal peak value of the load voltage of a bridge rectifier is equal to the peak value of the secondary voltage. **(4-4)** 8. ___
9. The bridge rectifier is the most popular rectifier design. **(4-4)** 9. ___
10. One way to reduce the ripple voltage is by reducing the discharging time constant of the capacitor-input filter. **(4-5)** 10. ___
11. Another way to reduce the ripple voltage is to use a full-wave rectifier or bridge rectifier instead of a half-wave rectifier. **(4-5)** 11. ___
12. If the filter's capacitance is decreased, then the ripple voltage will increase. **(4-5)** 12. ___
13. PIV stands for the peak intense voltage. **(4-6)** 13. ___
14. In a bridge rectifier the diode's PIV is equal to the peak value of the secondary voltage. **(4-6)** 14. ___
15. In a bridge rectifier the dc load current can exceed the maximum dc current rating of the diodes without endangering the diode of burning out. **(4-6)** 15. ___
16. The initial charging current of an uncharged capacitor filter is usually quite small. **(4-7)** 16. ___
17. If the load resistor of a capacitor-input filter is opened when used with a bridge rectifier, then the output would become a pure dc voltage. **(4-8)** 17. ___

18. If the capacitor is opened in a capacitor-input filter when used with a bridge rectifier, then the output would be a full-wave rectified voltage. (4-8)   18. ___
19. On the data sheet of the 1N4001, the peak repetitive reverse voltage is the same as the PIV. (4-9)   19. ___
20. On the data sheet of the 1N4001, I_0 is the designated symbol for the maximum allowable average rectified forward current. (4-9)   20. ___

II. COMPLETION

Complete each of the following.

Answer

1. In a step-down transformer the current is stepped ____. (4-1)   1. _____
2. In a step-up transformer the voltage is stepped ____. (4-1)   2. _____
3. The average value of a half-wave rectified signal is ____ percent of its peak value. (4-2)   3. _____
4. The average value of a rectified signal is called the ____ value. (4-2)   4. _____
5. The input frequency of a full-wave rectifier is equal to ____ times the output frequency. (4-3)   5. _____
6. A full-wave rectifier is characterized by having the center tap of the secondary windings ____. (4-3)   6. _____
7. The number of diodes that are used in a bridge rectifier is ____. (4-4)   7. _____
8. In a bridge rectifier, the number of conducting diodes that are in series with the load resistor during each half cycle is ____. (4-4)   8. _____
9. The discharging time constant of a capacitor-input filter is equal to the product of the load resistor and the ____. (4-5)   9. _____
10. If the ripple frequency of a capacitor-input filter was increased, the peak-to-peak value of the ripple voltage would ____. (4-5)   10. _____
11. In most circuits encountered, the ripple voltage is less than ____ percent of the dc load voltage. (4-5)   11. _____
12. If the load resistance of a capacitor-input filter is increased, then the ripple voltage will ____. (4-5)   12. _____
13. In a bridge rectifier, the diode's peak inverse voltage is equal to the peak value of the transformer's ____ voltage. (4-6)   13. _____
14. In a bridge rectifier, the dc current through the conducting diodes is ____ percent of the dc load current. (4-6)   14. _____
15. The sudden gush of current when the power is turned on, due to an initially uncharged filter capacitor, is called the ____ current. (4-7)   15. _____
16. If a diode is opened in a bridge rectifier with a capacitor-input filter, then the line frequency would equal the ____ frequency. (4-8)   16. _____
17. If the capacitor is opened in a capacitor-input filter with a bridge rectifier, then the dc load voltage would be equal to the ____ value of the unfiltered full-wave signal. (4-8)   17. _____
18. The electronic instrument that is commonly used to measure the ripple voltage is an ____. (4-8)   18. _____
19. On the data sheet of a 1N4001, the symbol for the maximum allowable peak repetitive inverse voltage is ____. (4-9)   19. _____
20. On the data sheet of a 1N4001, the symbol for the maximum allowable surge current is ____. (4-9)   20. _____

III. HIGHLIGHTS OF CIRCUITS

A. The Input Transformer

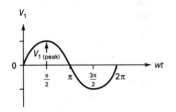 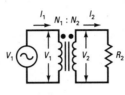 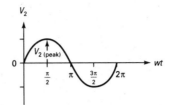

Figure 4-1

Formulas

$$V_p = V_{peak} = 1.414 V_{rms} \qquad *V_2 = \frac{N_2}{N_1} V_1 \qquad *I_2 = \frac{N_1}{N_2} I_1$$

$$I_p = I_{peak} = 1.414 I_{rms} \qquad \text{* Either rms or peak value.}$$

Ideal Rectified Output

B. Half-Wave Rectifier

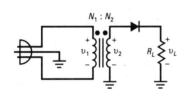

Figure 4-2

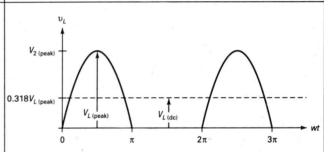

Figure 4-5

C. Full-Wave Rectifier

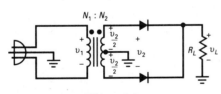

Figure 4-3

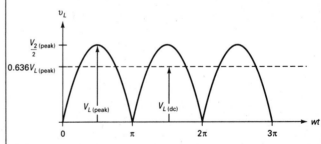

Figure 4-6

D. Bridge Rectifier

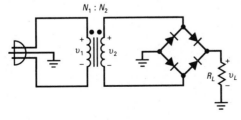

Figure 4-4

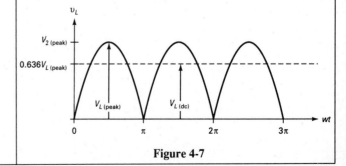

Figure 4-7

		Half-Wave	Full-Wave	Bridge
$V_{L(peak)}$	Peak load voltage	$V_{2(peak)}$	$V_{2(peak)}/2$	$V_{2(peak)}$
$V_{L(dc)}$	DC load voltage	$0.318 V_{2(peak)}$	$0.636 V_{2(peak)}/2$	$0.636 V_{2(peak)}$
$I_{L(dc)}$	DC load current	$V_{L(dc)}/R_L$	$V_{L(dc)}/R_L$	$V_{L(dc)}/R_L$
$I_{D(dc)}$	DC diode current	$I_{L(dc)}$	$0.5 I_{L(dc)}$	$0.5 I_{L(dc)}$
PIV	Peak inverse voltage	$V_{2(peak)}$	$V_{2(peak)}$	$V_{2(peak)}$
f_{out}	Ripple frequency	f_{in}	$2 f_{in}$	$2 f_{in}$
$V_{R(rms)}$	RMS ripple voltage†	$0.385 V_{2(peak)}$	$0.305 V_{2(peak)}/2$	$0.305 V_{2(peak)}$
$V_{L(rms)}$	RMS load voltage‡	$0.5 V_{2(peak)}$	$0.707 V_{2(peak)}/2$	$0.707 V_{2(peak)}$
% ripple	Percent ripple factor§	121%	48%	48%

Table 4-1 Ideal Rectifiers

* The load voltage is made up of a ripple voltage and a dc voltage.
† $V_{R(rms)}$ RMS value of the ripple voltage is derived from calculus.
‡ $V_{L(rms)} = \sqrt{V_{L(dc)}^2 + V_{R(rms)}^2}$ Derived from $P_L = P_{L(dc)} + P_{L(ripple)}$.
§ % ripple $= \dfrac{V_{R(rms)}}{V_{L(dc)}} \times 100\%$ By definition—figure of merit.

E. The Bridge Rectifier with a Capacitor-Input Filter

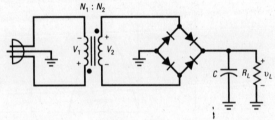

Figure 4-8

1. Output Waveform without Load

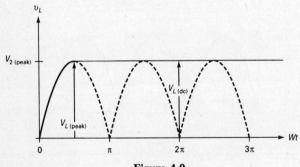

Figure 4-9

2. Output Waveform with Load

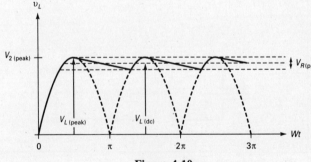

Figure 4-10

		Without Load	With Load
$V_{L(peak)}$	Peak load voltage	$V_{2(peak)}$	$V_{2(peak)}$
$V_{L(dc)}$	DC load voltage	$V_{2(peak)}$	$V_{2(peak)} - V_{R(pp)}/2$
$I_{L(dc)}$	DC load current	$V_{L(dc)}/R_L$	$V_{L(dc)}/R_L$
f_{out}	Ripple frequency	0	$2f_{in}$
$V_{R(pp)}$	Ripple voltage peak to peak	0	$I_{L(dc)}/f_{out}C$*
$V_{R(rms)}$	RMS ripple voltage	0	$V_{R(pp)}/2\sqrt{3}$†
$V_{L(rms)}$	RMS load voltage	$V_{2(peak)}$	$\sqrt{V_{L(dc)}^2 + V_{R(rms)}^2}$‡
% ripple	Percent ripple factor	0	$(V_{R(rms)}/V_{L(dc)}) \times 100\%$§

Table 4-2 An Ideal Bridge Rectifier with a Capacitor-Input Filter

* $V_{R(pp)} = I_{L(dc)}/f_{out}C$ Derived mathematically.
† $V_{R(rms)} = V_{R(pp)}/2\sqrt{3}$ Assume ripple is a triangular wave.
‡ $V_{L(rms)} \approx V_{dc}$.
§ % ripple $= \dfrac{1}{2\sqrt{3}f_{out}CR_L} \times 100\%$ By substitution.

Note: For accuracy, use the four above relationships only when the ripple voltage is less than 0.1 of $V_{L(peak)}$—light load condition.

IV. ILLUSTRATIVE AND PRACTICE PROBLEMS

Illustrative Problem 1. Find the dc load voltage and current for the following circuit:

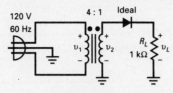

Figure 4-11

Steps	Comments
1. Find $V_{1(peak)}$. $V_{1(peak)} = 1.414\,(120\text{ V}) = 170\text{ V}$	Convert the rms value to the peak value. $V_{1(peak)} = 1.414\,V_{1(rms)}$
2. Find $V_{2(peak)}$. $V_{2(peak)} = \frac{1}{4}\,170\text{ V} = 42.5\text{ V}$	Use the transformer turns ratio to determine the secondary voltage. $V_{2(peak)} = \frac{N_2}{N_1}\,V_{1(peak)}$
3. Find $V_{L(peak)}$. $V_{L(peak)} = 42.5\text{ V}$	Diode is ideal. $V_{L(peak)} = V_{2(peak)}$
4. Find $V_{L(dc)}$. $V_{L(dc)} = 0.318\,(42.5\text{ V}) = 13.5\text{ V}$	The dc value equals the average value of the half-wave rectified waveform. $V_{L(dc)} = 0.318\,V_{out(peak)}$
5. Find $I_{L(dc)}$. $I_{L(dc)} = \frac{13.5\text{ V}}{1\text{ k}\Omega} = 13.5\text{ mA}$	Ohm's law. $I_{L(dc)} = \frac{V_{L(dc)}}{R_L}$

Practice Problem 1. Find the dc load voltage and current for the following circuit:

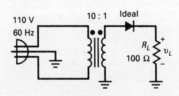

Figure 4-12

Answers: $V_{L(dc)} = 4.96\text{ V}$; $I_{L(dc)} = 49.6\text{ mA}$.

Illustrative Problem 2. The transformer's secondary peak voltage is 50 V. Determine how much power the load resistor is dissipating.

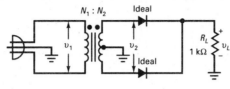

Figure 4-13

Steps	Comments
1. Find $V_{L(peak)}$. $$V_{L(peak)} = \frac{50\text{ V}}{2} = 25\text{ V}$$	Full-wave rectifier with ideal diodes. $$V_{L(peak)} = \frac{V_{2(peak)}}{2}$$
2. Find $V_{L(dc)}$. $$V_{L(dc)} = 0.636\,(25\text{ V}) = 15.9\text{ V}$$	The dc value equals the average value of the full-wave rectified waveform. $$V_{L(dc)} = 0.636 V_{L(peak)}$$
3. Find $V_{R(rms)}$. $$V_{R(rms)} = 0.305\,(25\text{ V}) = 7.63\text{ V}$$	The ripple's rms value for a full-wave rectified waveform. $$V_{R(rms)} = 0.305 V_{L(peak)}$$
4. Find $V_{L(rms)}$. $$V_{L(rms)} = \sqrt{(15.9\text{ V})^2 + (7.63\text{ V})^2} = 17.6\text{ V}$$	The rms value of the total load voltage. $$V_{L(rms)} = \sqrt{V_{L(dc)}^2 + V_{R(rms)}^2}$$
5. Find P_L. $$P_L = \frac{(17.6\text{ V})^2}{1\text{ k}\Omega} = 310\text{ mW}$$	The power dissipated by the load resistor. Power formula. $$P_L = \frac{V_{L(rms)}^2}{R_L}$$

Practice Problem 2. The transformer's secondary peak voltage is 30 V. Determine how much power the load resistor is dissipating.

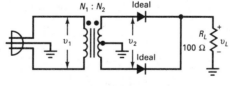

Figure 4-14

Answer: $P_L = 1.12$ W.

Illustrative Problem 3. Find the dc load voltage, the percent ripple factor, and the power dissipated by the load for the following circuit:

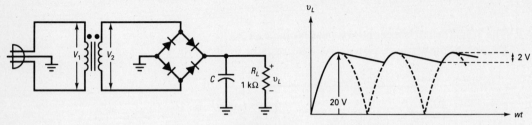

Figure 4-15

Steps	Comments
1. Find $V_{L(dc)}$. $V_{L(dc)} = 20\text{ V} - \dfrac{2\text{ V}}{2} = 19\text{ V}$	The dc value is equal to the average value. $V_{L(dc)} = V_{L(peak)} - \dfrac{V_{R(pp)}}{2}$
2. Find $V_{R(rms)}$. $V_{R(rms)} = \dfrac{2\text{ V}}{2\sqrt{3}} = 0.577\text{ V}$	Since $V_{R(pp)} \leq 0.1 V_{L(peak)}$, you can assume that the ripple waveform is triangular. $V_{R(rms)} = \dfrac{V_{R(pp)}}{2\sqrt{3}}$
3. Find percent ripple. $\%\text{ ripple} = \dfrac{0.577\text{ V}}{19\text{ V}} \times 100\% = 3.03\%$	Definition of percent ripple factor. $\%\text{ ripple} = \dfrac{V_{R(rms)}}{V_{L(dc)}} \times 100\%$
4. Find $V_{L(rms)}$. $V_{L(rms)} = \sqrt{(19\text{ V})^2 + (0.577\text{ V})^2} = 19.009\text{ V}$	The rms value of the total load voltage. $V_{L(rms)} = \sqrt{V_{L(dc)}^2 + V_{R(rms)}^2}$ Note that $V_{L(rms)} \approx V_{L(dc)}$.
5. Find P_L. $P_L = \dfrac{(19\text{ V})^2}{1\text{ k}\Omega} = 361\text{ mW}$	The power dissipated by the load resistor. Power formula. $P_L = \dfrac{V_{L(rms)}^2}{R_L}$

Practice Problem 3. Find the dc load voltage, the percent ripple factor, and the power dissipated by the load for the following circuit.

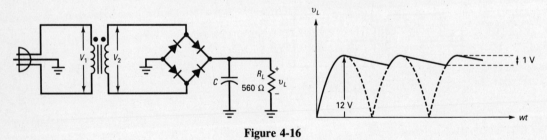

Figure 4-16

Answers: $V_{L(dc)} = 11.5\text{ V}$, % ripple = 2.51%, $P_L = 236\text{ mW}$.

V. PROBLEMS

Sec. 4-1 The Input Transformer

4-1. In the circuit of Fig. 4-17, the reading of the ac voltmeter is 15 V rms. Find the following:
   a. The peak value of the secondary voltage.
   b. The rms value of the primary voltage.
   c. The peak value of the secondary current.
   d. The rms value of the primary current.
   e. The power dissipated by the load resistor.

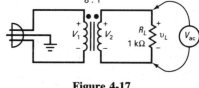
Figure 4-17

4-2. A step-up transformer has a primary voltage of 120 V rms and a secondary voltage of 679 V peak. What is the turns ratio?

4-3. In the circuit of Fig. 4-18, the reading of the ac ammeter is 0.02 A rms and the reading of the ac voltmeter is 110 V rms. Find the following:
   a. The peak value of the secondary voltage.
   b. The rms value of the load current.
   c. The value of the load resistor.
   d. The power dissipated by the load resistor.

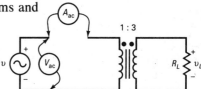
Figure 4-18

Sec. 4-2 The Half-Wave Rectifier

4-4. For the half-wave rectifier voltage waveform of Fig. 4-19, find the following (see "Highlights of Circuits"):
   a. The dc value of the voltage.
   b. The rms value of the ripple voltage.
   c. The rms value of the total voltage.
   d. The frequency of the ripple voltage.

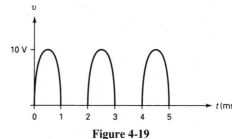
Figure 4-19

4-5. In the circuit of Fig. 4-20, the reading of the ac voltmeter is 20 V rms. Find the following:
   a. The peak value of the load voltage.
   b. The dc value of the load voltage.
   c. The dc value of the load current.

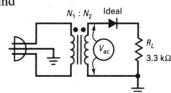
Figure 4-20

4-6. In the circuit of Fig. 4-21, find the peak value of the load voltage for the following conditions:
   a. An ideal diode.
   b. For the second approximation (silicon diode).

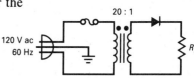
Figure 4-21

4-7. In the circuit of Fig. 4-22, find the following (see "Highlights of Circuits"):
   a. $V_{L(peak)}$
   b. $V_{L(dc)}$
   c. $V_{R(rms)}$
   d. $V_{L(rms)}$
   e. $I_{L(dc)}$
   f. P_L
   g. Percent ripple
   h. f_{out}

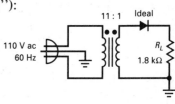
Figure 4-22

Sec. 4-3 The Full-Wave Rectifier

4-8. For the full-wave rectifier voltage waveform of Fig. 4-23, find the following (see "Highlights of Circuits"):
  a. The dc value of the voltage.
  b. The rms value of the ripple voltage.
  c. The rms value of the total voltage.
  d. The frequency of the ripple voltage.

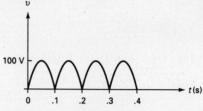

Figure 4-23

4-9. In the circuit of Fig. 4-24, the reading of the ac voltmeter is 30 V rms. Find the following:
  a. The peak value of the load voltage.
  b. The dc value of the load voltage.
  c. The dc value of the load current.

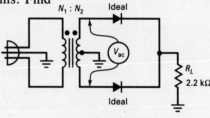

Figure 4-24

4-10. In the circuit of Fig. 4-25, find the peak value of the load voltage for the following conditions:
  a. Ideal diodes.
  b. For the second approximation (silicon diodes).

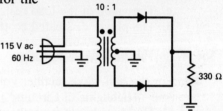

Figure 4-25

4-11. In the circuit of Fig. 4-26, find the following (see "Highlights of Circuits"):
  a. $V_{L(peak)}$ b. $V_{L(dc)}$
  c. $V_{R(rms)}$ d. $V_{L(rms)}$
  e. $I_{L(dc)}$ f. P_L
  g. Percent ripple   h. f_{out}

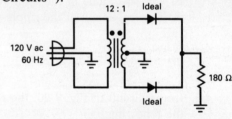

Figure 4-26

Sec. 4-4 The Bridge Rectifier

4-12. In the circuit of Fig. 4-27, the reading of the ac voltmeter is 30 V rms. Find the following:
  a. The peak value of the load voltage.
  b. The dc value of the load voltage.
  c. The dc value of the load current.

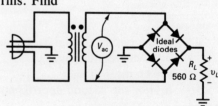

Figure 4-27

4-13. In the circuit of Fig. 4-28, find the peak value of the load voltage for the following conditions:
  a. Ideal diodes.
  b. For the second approximation (silicon diodes).

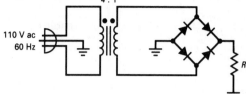

Figure 4-28

4-14. In the circuit of Fig. 4-29, find the following (see "Highlights of Circuits"):
  a. $V_{L(peak)}$ b. $V_{L(dc)}$
  c. $V_{R(rms)}$ d. $V_{L(rms)}$
  e. $I_{L(dc)}$ f. P_L
  g. Percent ripple   h. f_{out}

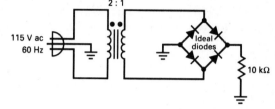

Figure 4-29

4-15. In the circuit of Fig. 4-30, the reading of the dc voltmeter is 12 V. Find the following:
  a. $V_{1(peak)}$
  b. $V_{L(rms)}$

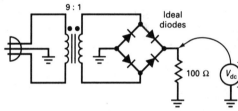

Figure 4-30

Sec. 4-5 The Capacitor-Input Filter

4-16. For the output voltage waveform of the capacitor-input filter shown in Fig. 4-31, find the following (see "Highlights of Circuits"):
  a. The peak-to-peak value of the ripple voltage.
  b. The dc value of the voltage.
  c. The rms value of the total voltage.
  d. Percent ripple.
  e. The ripple frequency.

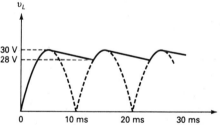

Figure 4-31

4-17. In the circuit of Fig. 4-32, the dc voltmeter reading is 15 V. Find the following (*Hint:* Assume that $V_{R(pp)} < 0.1\ V_{L(peak)}$) (also see "Highlights of Circuits"):
  a. $I_{L(dc)}$
  b. $V_{R(pp)}$
  c. $V_{L(peak)}$
  d. $V_{R(rms)}$
  e. Percent ripple
  f. P_L

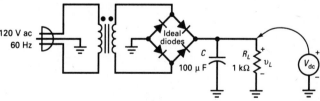

Figure 4-32

4-18. In the circuit of Fig. 4-33, the reading of the ac voltmeter is 28.3 V rms and the reading of the dc voltmeter is 39 V. Find the following (see "Highlights of Circuits"):
  a. $V_{R(pp)}$
  b. $V_{R(rms)}$
  c. Percent ripple
  d. The value of the capacitor.
  e. P_L

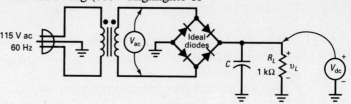

Figure 4-33

Sec. 4-6 Calculating Other Quantities

4-19. Determine the value of the dc diode's current and the peak inverse voltage for the circuit of Prob. 4-11.

4-20. Determine the value of the dc diode's current and the peak inverse voltage for the circuit of Prob. 4-14.

CHAPTER 5

SPECIAL-PURPOSE DIODES

Study Chap. 5 in *Electronic Principles*.

I. TRUE / FALSE

Answer true (T) or false (F) to each of the following statements.

Answer

1. A zener diode is designed by the manufacturer to operate in its forward region. (5-1) 1. ___
2. In a zener regulator a series resistor is always used to limit the zener current to less than its maximum current rating. (5-1) 2. ___
3. The manufacturer varies the doping level of silicon diodes to produce zener diodes with breakdown voltages from about 2 to 200 V. (5-1) 3. ___
4. In a loaded zener regulator, the zener diode and the load resistor are in series. (5-2) 4. ___
5. The amount of ripple voltage across the load resistor of a zener regulator depends on the value of the zener resistance. (5-2) 5. ___
6. The temperature coefficient of a zener diode is always negative. (5-2) 6. ___
7. For most of the commercially available LEDs, the typical voltage drop is from 1.5 to 2.5 V for currents between 10 and 50 mA. (5-3) 7. ___
8. To utilize the properties of photodiodes best, they should be forward-biased. (5-3) 8. ___
9. The key advantage of an optocoupler is the electrical isolation between the input and output circuits. (5-3) 9. ___
10. A Schottky diode has a barrier potential of 0.7 V when it is forward-biased. (5-4) 10. ___
11. The most important application of Schottky diodes is in digital computers because of their extremely fast reverse recovery time. (5-4) 11. ___
12. The forward-biased voltage controls the capacitance of the varactor. (5-5) 12. ___
13. A varactor can be connected in parallel with an inductor to produce a resonant circuit. (5-5) 13. ___
14. The varistor is a device that is used for line filtering. (5-6) 14. ___
15. The varistor works best in eliminating voltage dips, rather than voltage spikes. (5-6) 15. ___
16. As long as the operating power is less than the rated power, the zener diode can operate in the breakdown region without being destroyed. (5-7) 16. ___

17. The maximum current rating of a zener diode is not related to its maximum power rating. (5-7)   17. ___
18. The derating factor of a zener diode is temperature dependent. (5-7)   18. ___
19. When troubleshooting, voltage measurements give clues that help isolate the trouble. (5-8)   19. ___
20. A voltage drop across a zener diode that exceeded its rated value would indicate that it is blown. (5-8)   20. ___

II. COMPLETION

Complete each of the following.

Answer

1. A zener diode is sometimes called a voltage-___ diode. (5-1)   1. ___
2. A zener diode operating in its breakdown region ideally acts like a ___. (5-1)   2. ___
3. Circuits that hold the load voltage almost constant despite large changes in line voltage and load resistance are called voltage ___. (5-1)   3. ___
4. In an unloaded zener regulator, the zener current is equal to the ___ current. (5-2)   4. ___
5. The effects of temperature change that cause zener voltage variations are listed on data sheets under the temperature ___. (5-2)   5. ___
6. Ideally, the zener voltage is constant and is equal to the load voltage; therefore, the zener regulator reduces any ripple voltage to ___. (5-2)   6. ___
7. A diode that has been optimized for its sensitivity to light is called a ___. (5-3)   7. ___
8. In a forward-biased LED, when the electrons fall from a high to a low energy level, they radiate energy in the form of ___. (5-3)   8. ___
9. The combination of an LED and a photodiode in a single package is called a ___. (5-3)   9. ___
10. A type of diode that has no depletion layer, which eliminates the stored charges at the junction, is called a ___ diode. (5-4)   10. ___
11. The time it takes to turn off a forward-biased diode is called the reverse ___ time. (5-4)   11. ___
12. The effect of how increases in frequency cause a diode to reach a point where it cannot turn off fast enough is known as ___ storage. (5-4)   12. ___
13. The varactor is also called a ___ diode. (5-5)   13. ___
14. At high frequencies, the varactor acts the same as a variable ___. (5-5)   14. ___
15. The varistor is also called a ___ suppressor. (5-6)   15. ___
16. The varistor behaves like two back-to-back ___ diodes. (5-6)   16. ___
17. What is shown on a data sheet that tells you how much you have to reduce the power rating of a device for lead temperature operations is called a ___ factor. (5-7)   17. ___
18. The zener resistance is also called the zener ___. (5-7)   18. ___
19. The suffix *A* or *B* that is used in conjunction with a zener identification number indicates the zener ___. (5-7)   19. ___
20. When a circuit is not working as it should, a troubleshooter usually starts measuring ___. (5-8)   20. ___

III. ILLUSTRATIVE AND PRACTICE PROBLEMS

Illustrative Problem 1. Find V_{out} and I_Z for the unloaded 10-V zener regulator that has an internal resistance of 5 Ω by using the ideal and second approximation methods.

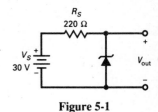

Figure 5-1

Steps	Comments
1. Find V_{out} and I_Z (ideal). Figure 5-2 $V_{out} = 10$ V $I_Z = \dfrac{30\text{ V} - 10\text{ V}}{220\ \Omega} = 91$ mA	a. Since the source voltage exceeds the zener voltage, the zener diode is operating in its breakdown region. b. For the ideal approximation, replace the zener diode with a battery whose voltage is the zener voltage (neglect R_Z). c. Solve for V_{out} and I_Z.
2. Find V_{out} and I_Z (second approximation). Figure 5-3 $I_Z = \dfrac{30\text{ V} - 10\text{ V}}{225\ \Omega} = 88.9$ mA $V_{out} = 10\text{ V} + 88.9\text{ mA}\ (5\ \Omega) = 10.44$ V	a. The zener is operating in its breakdown region. b. For the second approximation, replace the zener diode with a battery whose voltage is the zener voltage in series with the zener's internal resistance. c. Solve for V_{out} and I_Z.

Practice Problem 1. Find V_{out} and I_Z for the unloaded 20-V zener regulator that has an internal resistance of 12 Ω by using the ideal and second approximation methods.

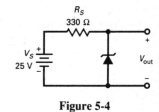

Figure 5-4

Answers:	V_{out}	I_Z
Ideal	20 V	15.2 mA
Second approximation	20.18 V	14.6 mA

Copyright © 1993 by the Glencoe Division of Macmillan/McGraw-Hill School Publishing Company. All rights reserved.

Illustrative Problem 2. For the loaded 12-V ideal zener regulator, find the following:
 a. V_L
 b. I_L
 c. I_S
 d. I_Z
 e. P_Z

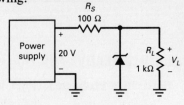

Figure 5-5

Steps	Comments
1. Find V_{TH} and compare it to V_Z. Figure 5-6 $V_{TH} = \dfrac{1 \text{ k}\Omega}{1.1 \text{ k}\Omega} 20 \text{ V} = 18.18 \text{ V}$ $V_Z = 12 \text{ V} \therefore V_{TH} > V_Z$	a. Calculate V_{TH} (open-circuit voltage with the zener diode removed). b. Is $V_{TH} > V_Z$? c. If V_{TH} exceeds the zener voltage, then the diode is operating in its breakdown region.
2. Find V_L, I_L, I_S, I_Z, and P_Z. Figure 5-7 $V_L = 12 \text{ V}$ $I_L = \dfrac{12 \text{ V}}{1 \text{ k}\Omega} = 12 \text{ mA}$ $I_S = \dfrac{20 \text{ V} - 12 \text{ V}}{100 \text{ }\Omega} = 80 \text{ mA}$ $I_Z = 80 \text{ mA} - 12 \text{ mA} = 68 \text{ mA}$ $P_Z = (12 \text{ V})(68 \text{ mA}) = 816 \text{ mW}$	a. Since V_{TH} exceeds V_Z, replace the ideal zener diode with a battery whose voltage is the zener voltage. b. Solve for V_L, I_L, I_S, I_Z, and P_Z. c. If V_{TH} does not exceed the zener voltage, the zener diode is off. Under this condition remove the zener diode from the circuit and solve for V_L, I_S, and I_L. I_Z and P_Z would be equal to zero and I_S would be equal to I_L.

Practice Problem 2. For the loaded 15-V ideal zener regulator, find the following:
 a. V_L
 b. I_L
 c. I_S
 d. I_Z
 e. P_Z

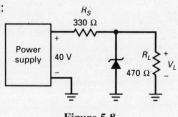

Figure 5-8

Answers:
 a. $V_L = 15 \text{ V}$
 b. $I_L = 31.9 \text{ mA}$
 c. $I_S = 75.8 \text{ mA}$
 d. $I_Z = 43.9 \text{ mA}$
 e. $P_Z = 659 \text{ mW}$

IV. PROBLEMS

Sec. 5-1 The Zener Diode

5-1. Find V_{out} and I_Z for the unloaded 25-V zener regulator that has an internal resistance of 10 Ω in Fig. 5-9. Use the ideal and second approximation methods.

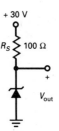

Figure 5-9

5-2. The circuit of Fig. 5-10 is an unloaded 15-V ideal zener regulator. Do the following:
  a. Complete Table 5-1 for the given conditions.
  b. Determine the range of values that the supply voltage may have in order that regulation is maintained.
  c. What is the purpose of R_S?

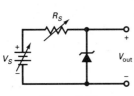

Figure 5-10

V_S, V	R_S, Ω	I_Z, mA	V_{out}, V
10	1 k		
10	100		
15	1 k		
15	100		
20	1 k		
20	100		
50	1 k		
50	100		

Table 5-1

5-3. When the current of a specific zener diode is increased from 20 to 30 mA, the zener voltage changes from 6.2 to 6.3 V. What is the resistance of the zener diode?

Sec. 5-2 The Loaded Zener Regulator

5-4. For the loaded 18-V ideal zener regulator in Fig. 5-11, find:
  a. V_L
  b. I_L
  c. I_S
  d. I_Z
  e. P_Z

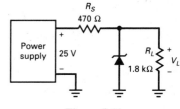

Figure 5-11

5-5. The circuit of Fig. 5-12 is a loaded 12-V ideal zener regulator. Do the following:
  a. Complete Table 5-2 for the given circuit conditions.
  b. Determine the range of values that the supply voltage may have in order to maintain the regulation.

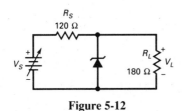

Figure 5-12

V_S, V	V_L, V	I_L, mA	I_S, mA	I_Z, mA
10				
20				
30				
40				
50				

Table 5-2

5-6. The circuit of Fig. 5-13 is a loaded 10-V ideal zener regulator. Do the following:
  a. Complete Table 5-3 for the given circuit conditions.
  b. Determine the range of values that the load may have in order to maintain the regulation.

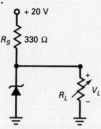

Figure 5-13

R_L, Ω	V_L, V	I_L, mA	I_S, mA	I_Z, mA
10				
100				
330				
470				
1 k				
10 k				

Table 5-3

5-7. In the circuit of Fig. 5-14, what value must R_S be adjusted to so that the zener current of the 10-V ideal zener diode is 50 mA?

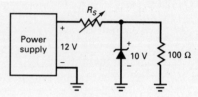

Figure 5-14

5-8. Determine the peak-to-peak value of the load's ripple voltage for the circuit of Fig. 5-11 and the voltage waveforms of Fig. 5-15, if the voltage of the zener regulator is 18 V, and the resistance of the zener diode is 5 Ω.

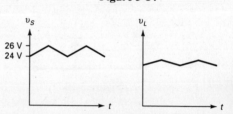

Figure 5-15

5-9. From the data obtained from the voltage measurements for the circuit of Fig. 5-16, determine the resistance of the 10-V zener diode.

Data: 1. $V_1 = 20$ V, $V_2 = 10.1$ V
2. $V_1 = 22$ V, $V_2 = 10.2$ V

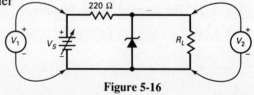

Figure 5-16

Sec. 5-3 Optoelectronics Devices

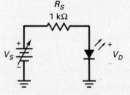

Figure 5-17

5-10. For the circuit of Fig. 5-17, do the following:
  a. Complete Table 5-4 for the given circuit conditions.
  b. What is the purpose of R_S?
  c. Determine the range of values that the supply voltage may be set to so that current will flow through the LED.

V_S, V	V_D, V	I_D, mA
1		
2		
5		
10		
20		

Table 5-4

Name ______________________________ Date ______________

5-11. For the circuit of Fig. 5-18, do the following:
a. Complete Table 5-5 for the given circuit conditions.
b. Determine the range of values that R_S must be to limit the current that flows through the LED from 10 to 18 mA.

Figure 5-18

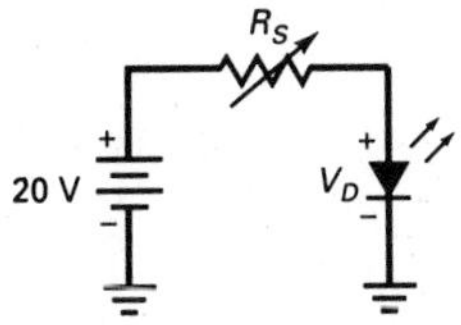

R_S, Ω	V_D, V	I_D, mA
150		
1 k		
1.2 k		
1.8 k		
10 k		

Table 5-5

5-12. For the circuit of Fig. 5-19, what value must R_S be adjusted to so that the LED current is 16 mA?

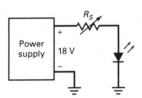

Figure 5-19

Sec. 5-7 Reading a Data Sheet

5-13. In the circuit of Fig. 5-20, is the maximum power rating of the IN961 zener diode being exceeded?

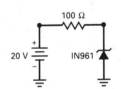

Figure 5-20

5-14. In the circuit of Fig. 5-21, is the maximum current rating of the IN968 zener diode being exceeded?

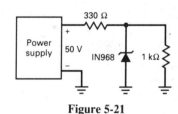

Figure 5-21

5-15. What is the maximum impedance of the 1N758 zener diode?

Sec. 5-8 Troubleshooting

5-16. The circuit of Fig. 5-22 is a loaded 5-V ideal zener regulator. Fill in the readings of the voltmeters in Table 5-6 for the given circuit conditions.

Trouble	Voltmeter Readings	
	V_1, V	V_2, V
Circuit OK		
Zener opened		
Zener shorted		
R_L opened		
R_L shorted		
Zener inserted with polarity reversed		
Power supply off		

Table 5-6

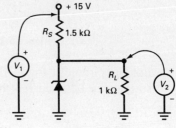

Figure 5-22

CHAPTER 6

BIPOLAR TRANSISTORS

Study Chap. 6 in *Electronic Principles*.

I. TRUE / FALSE

Answer true (T) or false (F) to each of the following statements.

Answer

1. The number of doped regions in a bipolar transistor is two. **(6-1)** 1. ___
2. For each depletion layer in a silicon bipolar transistor, the barrier potential at 25° C is approximately 0.7 V. **(6-1)** 2. ___
3. The collector diode of a bipolar transistor is formed by the collector and the base. **(6-1)** 3. ___
4. The two types of bipolar transistors that are manufactured are called *npn* and *pnp* devices. **(6-1)** 4. ___
5. The emitter of a transistor is lightly doped. **(6-2)** 5. ___
6. The base of a transistor is lightly doped and very thin. **(6-2)** 6. ___
7. In order for a transistor to operate properly, the emitter diode must be reverse-biased. **(6-2)** 7. ___
8. In most transistors, more than 95 percent of the emitter electrons flows to the collector and less than 5 percent flows out the external base lead. **(6-2)** 8. ___
9. The collector current is the sum of the emitter current and the base current. **(6-3)** 9. ___
10. In low power transistors, the base current is usually less than 1 percent of the collector current. **(6-3)** 10. ___
11. The base current controls the collector current. **(6-4)** 11. ___
12. The supply voltage of the collector must forward-bias the collector diode or else the transistor will not work properly. **(6-4)** 12. ___
13. The base curve of a transistor looks like an ordinary rectifier diode curve. **(6-5)** 13. ___
14. A transistor is not intended for operation in the breakdown region. **(6-6)** 14. ___
15. The transistor has four distinct operating regions. **(6-6)** 15. ___
16. When the transistor is operating in the saturation region, the collector diode is reverse-biased. **(6-6)** 16. ___
17. The horizontal axis of the transistor's collector curve represents the transistor's collector-emitter voltage. **(6-6)** 17. ___
18. The transistor power is equal to the collector-emitter voltage times the collector current. **(6-6)** 18. ___

19. For the second approximation, the base-emitter diode has no barrier potential and no bulk resistance. (6-7) 19. ___
20. The second approximation is a good compromise for troubleshooting and design. (6-7) 20. ___
21. The base-emitter part of an ideal transistor acts like an ideal diode. (6-7) 21. ___
22. The collector-emitter part of an ideal transistor acts like a constant current source. (6-7) 22. ___
23. The V_{CB} rating for the 2N3904 stands for the maximum operating reverse voltage from the collector to base. (6-8) 23. ___
24. On the data sheet of the 2N3904, the current gain is symbolized by β_{dc}. (6-8) 24. ___

II. COMPLETION

Complete each of the following statements.

Answer

1. The number of depletion layers in a bipolar transistor is ___. (6-1) 1. ___
2. The emitter diode of a bipolar transistor is formed by the emitter and the ___. (6-1) 2. ___
3. If the emitter is *p*-type material, then the base must be ___-type material. (6-1) 3. ___
4. If the collector is *n*-type material, then the emitter must be ___-type material. (6-1) 4. ___
5. An unbiased transistor is like two back-to-back ___. (6-2) 5. ___
6. In order for a transistor to operate properly the collector diode must be ___-biased. (6-2) 6. ___
7. The collector is so named because it collects electrons from the ___. (6-2) 7. ___
8. The emitter is so named because it emits electrons to the ___. (6-2) 8. ___
9. The largest of the transistor's current is the ___ current. (6-3) 9. ___
10. The collector current is approximately equal to the ___ current. (6-3) 10. ___
11. The current gain of a transistor is defined as the collector current divided by the ___ current. (6-3) 11. ___
12. The voltage between the collector and the emitter is symbolized by ___. (6-4) 12. ___
13. The base supply voltage and the base current-limiting resistance directly controls the ___ current. (6-4) 13. ___
14. The base-emitter part of a transistor behaves like a ___. (6-5) 14. ___
15. When the base current is zero, the transistor is operating in the ___ region. (6-6) 15. ___
16. When the collector-to-emitter voltage is between 0 and 1 V, the transistor is operating in the ___ region. (6-6) 16. ___
17. The region that represents normal operation of the transistor and is characterized by horizontal lines on the collector curve is called the ___ region. (6-6) 17. ___
18. The vertical axis of the transistor's collector curve represents the transistor's ___ current. (6-6) 18. ___
19. Transistors that are used as amplifiers operate in their ___ region. (6-6) 19. ___
20. For the third approximation of a silicon transistor, V_{BE} may be greater than 0.7 V because of the emitter ___ resistance. (6-7) 20. ___
21. For the third approximation, V_{CE} may be greater than zero when the transistor is operating in the ___ region. (6-7) 21. ___
22. The V_{CEO} rating for the 2N3904 stands for the maximum operating voltage from the collector to emitter with the base ___. (6-8) 22. ___
23. A way to increase the power rating of a power transistor is to attach the transistor's metal case to a ___ sink. (6-8) 23. ___
24. The system of analysis that uses h_{FE} as the symbol for current gain is called the ___ parameters. (6-8) 24. ___

III. ILLUSTRATIVE AND PRACTICE PROBLEMS

Illustrative Problem 1. For the circuit in Fig. 6-1, find the following:

a. I_B
b. I_C
c. I_E
d. V_C
e. V_{CE}
f. P_D

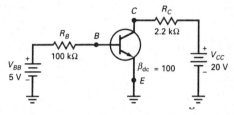

Figure 6-1

Steps	Comments
1. Find I_B. $$I_B = \frac{5\text{ V} - 0.7\text{ V}}{100\text{ k}\Omega} = 43\text{ μA}$$	a. $V_{BE} = 0.7$ V for a silicon transistor. b. Since $V_{BB} > V_{BE}$, the transistor is on. c. Find I_B by applying Ohm's law. $$I_B = \frac{V_{BB} - V_{BE}}{R_B}$$
2. Find I_C. $$I_C = (100)(43\text{ μA}) = 4.3\text{ mA}$$	Due to transistor action: $$I_C = \beta_{dc} I_B$$
3. Find I_E. $$I_E = 4.3\text{ mA} + 43\text{ μA} = 4.343\text{ mA}$$	The emitter current is equal to the sum of the collector and base currents: $$I_E = I_C + I_B$$
4. Find V_C. $$V_C = 20\text{ V} - (4.3\text{ mA})(2.2\text{ k}\Omega) = 10.5\text{ V}$$	The voltage at the collector is equal to the collector supply voltage minus the drop across the collector resistor. $$V_C = V_{CC} - I_C R_C$$
5. Find V_{CE}. $$V_{CE} = 10.5\text{ V} - 0\text{ V} = 10.5\text{ V}$$	V_{CE} is equal to the collector voltage minus the emitter voltage. $$V_{CE} = V_C - V_E$$
6. Find P_D. $$P_D = (4.3\text{ mA})(10.5\text{ V}) = 45.2\text{ mW}$$	Power dissipated by the transistor: $$P_D = I_C V_{CE}$$

Practice Problem 1. For the circuit in Fig. 6-2, find the following:

a. I_B
b. I_C
c. I_E
d. V_C
e. V_{CE}
f. P_D

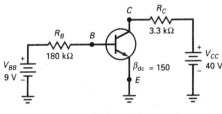

Figure 6-2

Answers:
a. $I_B = 46.1$ μA
b. $I_C = 6.92$ mA
c. $I_E = 6.96$ mA
d. $V_C = 17.2$ V
e. $V_{CE} = 17.2$ V
f. $P_D = 119$ mW

Illustrative Problem 2. For the circuit in Fig. 6-3, determine the value of the transistor's β_{dc} from the voltmeter readings.

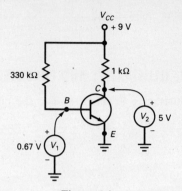

Figure 6-3

Steps	Comments
1. Find I_B. $$I_B = \frac{9\text{ V} - 0.67\text{ V}}{330\text{ k}\Omega} = 25.2\text{ }\mu\text{A}$$	Find I_B by applying Ohm's law. $$I_B = \frac{V_{CC} - V_1}{R_B}$$
2. Find I_C. $$I_C = \frac{9\text{ V} - 5\text{ V}}{1\text{ k}\Omega} = 4\text{ mA}$$	Find I_C by applying Ohm's law. $$I_C = \frac{V_{CC} - V_2}{R_C}$$
3. Find β_{dc}. $$\beta_{dc} = \frac{4\text{ mA}}{25.2\text{ }\mu\text{A}} = 159$$	By definition: $$\beta_{dc} = \frac{I_C}{I_B}$$

Practice Problem 2. For the circuit in Fig. 6-4, determine the value of the transistor's β_{dc} from the voltmeter readings.

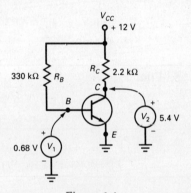

Figure 6-4

Answer: $\beta_{dc} = 87.5$.

IV. PROBLEMS

Sec. 6-3 Transistor Currents

6-1. In the circuit of Fig. 6-5, the transistor has a base current of 50 μA and an emitter current of 3 mA. What is the collector current?

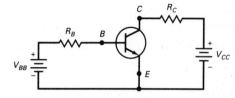

Figure 6-5

6-2. In the circuit of Fig. 6-6, the transistor has a base current of 40 μA and a collector current of 10 mA. What is the current gain of the transistor?

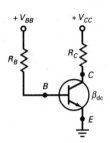

Figure 6-6

6-3. In the circuit of Fig. 6-7, the transistor has a current gain of 100 and a collector current of 6 mA. What are the emitter and base currents?

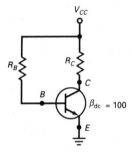

Figure 6-7

Sec. 6-5 Base Curve

6-4. For the circuit of Fig. 6-8, do the following:
  a. Complete Table 6-1 for the given circuit conditions.
  b. Determine the range of values that the base supply voltage must be in order for the base current to flow.
  c. What was the effect on the transistor's base current when only the base supply voltage was increased?
  d. What was the effect on the transistor's base current when only R_B was increased?

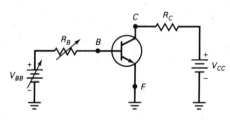

Figure 6-8

V_{BB}, V	R_B, Ω	I_B, μA
0.5	100 k	
0.5	1 M	
5	100 k	
5	1 M	
10	100 k	
10	1 M	

Table 6-1

6-5. For the circuit of Fig. 6-9, when V_{BB} is 10 V, what value must R_B be adjusted to so that the base current is 10 μA? (This is a simpler way to draw a transistor circuit.)

6-6. For the circuit of Fig. 6-9, when V_{BB} is 20 V, what value must R_B be adjusted to so that the base current is 10 μA?

6-7. What is the base current in the circuit of Fig. 6-10? (When $V_{BB} = V_{CC}$, the circuit can be drawn as shown.)

Sec. 6-6 Collector Curves

6-8. For the circuit of Fig. 6-11, find the following:
    a. I_B b. I_C c. I_E
    d. V_B e. V_E f. V_{BE}
    g. V_C h. V_{CE} i. P_D

6-9. For the circuit of Fig. 6-12, find the following:
    a. I_C
    b. V_{CE}
    c. P_D

6-10. For the circuit in Fig. 6-13, determine the value of the transistor's β_{dc} from the voltmeter readings.

6-11. For the circuit of Fig. 6-14, do the following:
    a. Complete Table 6-2 for the given circuit conditions.

R_C, kΩ	β_{dc}	I_B, μA	I_C, mA	V_{CE}, V
1	100			
1	150			
1.8	100			

Table 6-2

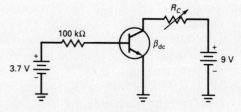

Figure 6-14

(Select either *increased*, *decreased*, or *stayed the same* as the correct answer to the following questions.)

    b. What happened to I_B when only β_{dc} was increased?
    c. What happened to I_B when only R_C was increased?
    d. What happened to I_C when only β_{dc} was increased?
    e. What happened to I_C when only R_C was increased?
    f. What happened to V_{CE} when only β_{dc} was increased?
    g. What happened to V_{CE} when only R_C was increased?

6-12. Identify each of the operating regions shown on the transistor's collector curves in Fig. 6-15.

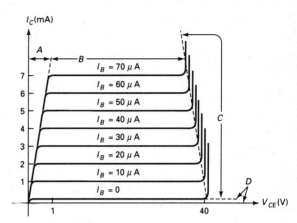

Figure 6-15

6-13. For the circuit of Fig. 6-16, do the following:
   a. Complete Table 6-3 for the given circuit conditions.
   b. In Table 6-3, indicate whether the transistor is operating in the saturation, active, or cutoff region.

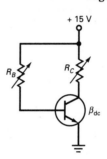

Figure 6-16

R_B, Ω	R_C, kΩ	β_{dc}	I_B, μA	I_C, mA	V_{CE}, V	Region of Operation
Opened	1	100				
220 k	1	100				
220 k	2.2	100				
220 k	1	225				
270 k	1	100				

Table 6-3

Sec. 6-7 Transistor Approximation

6-14. For the circuit of Fig. 6-17, find the following (give answers for the ideal and the second approximation):
   a. I_C b. V_{CE} c. P_D

6-15. For the circuit of Fig. 6-18, find the following (give answers for the ideal and the second approximation):
   a. I_C b. V_{CE} c. P_D

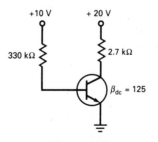

Figure 6-17

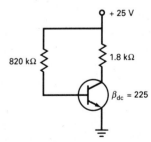

Figure 6-18

Sec. 6-8 Reading Data Sheets

6-16. The value of a measured β_{dc} of a 2N3904 is 50 for a collector current of 10 mA and a collector-emitter voltage of 1 V. Does this measured value fall within the manufacturer's specification for this device?

6-17. In the circuit of Fig. 6-19, is the 2N3904 transistor in danger of overheating?

6-18. In the circuit of Fig. 6-20, is the collector current of the 2N3904 exceeding its maximum rating?

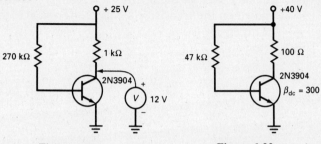

Figure 6-19   Figure 6-20

Sec. 6-9 Troubleshooting

6-19. For the circuit of Fig. 6-21, fill in the readings of the voltmeters in Table 6-4 for the given circuit conditions.

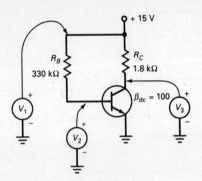

Figure 6-21

Trouble	Voltmeter Readings		
	V_1, V	V_2, V	V_3, V
Circuit OK			
Power supply off			
No base current			
R_B opened			
R_C shorted			
R_C opened			

Table 6-4

CHAPTER 7
TRANSISTOR FUNDAMENTALS

Study Chap. 7 in *Electronic Principles*.

I. TRUE / FALSE

Answer true (T) or false (F) to each of the following statements.

Answer

1. The current gain of a transistor is not affected by changes in the collector current. (7-1)   1. ___
2. The current gain of a transistor is affected by changes in the transistor's junction temperature. (7-1)   2. ___
3. The load line drawn on the collector curves is a visual summary of all the possible operating points that the circuit may have. (7-2)   3. ___
4. The saturation point is the point where the load line intercepts the saturation region on the collector curves. (7-2)   4. ___
5. In the base-biased circuit, the saturated value of the collector current is equal to the collector voltage divided by the collector resistance. (7-2)   5. ___
6. In a base-biased circuit, the cutoff voltage is equal to the collector supply voltage. (7-2)   6. ___
7. Every transistor circuit has a load line. (7-3)   7. ___
8. In order to draw a load line on the collector curves, you first have to determine the transistor's saturation current and cutoff voltage. (7-3)   8. ___
9. The mathematical proof of the load line is based on the concept of a linear equation and its graph. (7-4)   9. ___
10. The collector current can be greater than the saturation current. (7-5)   10. ___
11. In most digital circuit designs, the low and high outputs are produced when the transistor switches operation from saturation to cutoff. (7-6)   11. ___
12. Amplifiers depend on transistor circuits whose Q points are immune to changes in current gain. (7-7)   12. ___
13. In an emitter-biased circuit, the emitter current depends on the transistor's current gain. (7-7)   13. ___
14. When calculating for the coordinates of the Q point, you can assume that the collector current is equal to the emitter current for any β_{dc}. (7-7)   14. ___
15. Normally, a troubleshooter does not measure the collector-emitter voltage directly with a voltmeter. (7-7)   15. ___

16. To change the LED current in a base-biased LED driver, you can either change the collector resistor or the collector supply voltage. (7-8)   16. ___

17. To change the LED current in an emitter-biased LED driver you can change the collector supply voltage. (7-8)   17. ___

18. Dependent variables of a circuit are often called circuit values. (7-9)   18. ___

19. To understand how a circuit works suggests that you can predict how voltages and currents will react to small changes in circuit values. (7-9)   19. ___

20. When testing a transistor with an ohmmeter, the resistance measurements between the collector and emitter should be low in both directions. (7-10)   20. ___

II. COMPLETION

Complete each of the following statements.

1. The symbol for current gain on the manufacturer's data sheet is _____. (7-1)   1. _____

2. When a transistor is replaced by a transistor of the same type, the current gain of the replacement transistor can vary greatly due to the manufacturers' _____ of that device. (7-1)   2. _____

3. The symbol for the maximum possible collector current due to circuit conditions is _____. (7-2)   3. _____

4. In order to find the saturation current the collector-emitter's terminals must be _____. (7-2)   4. _____

5. The symbol for the transistor's cutoff voltage is _____. (7-2)   5. _____

6. In order to find the transistor's cutoff voltage the collector-emitter's terminals must be _____. (7-2)   6. _____

7. The operating point on the transistor's collector curves is often called the _____ point. (7-3)   7. _____

8. The symbol that is used on the collector curves to show the operating point is _____. (7-3)   8. _____

9. The concept of a load line can be proved mathematically and _____. (7-4)   9. _____

10. When using the technique *reductio ad absurdum*, you assume the transistor is operating in the active region and see if a _____ arises. (7-5)   10. _____

11. When the saturation current gain is much smaller than the active region current gain, the transistor is in _____ saturation. (7-5)   11. _____

12. When the saturation current gain is approximately a little less than the active region current gain, the transistor is in _____ saturation. (7-5)   12. _____

13. Digital circuits whose Q point switches between only two points on the load line are often called _____ circuits. (7-6)   13. _____

14. In an emitter bias circuit, the only resistor that affects the emitter current is the _____ resistance. (7-7)   14. _____

15. In an emitter bias circuit, the only supply voltage that affects the emitter current is the _____ supply voltage. (7-7)   15. _____

16. Base-biased circuits are normally designed to switch between saturation and cutoff, whereas emitter-biased circuits are usually designed to operate in the _____ region. (7-8)   16. _____

17. When the LED is on in a base-biased LED driver circuit, the transistor is in _____ saturation. (7-8)   17. _____

18. In an emitter-biased circuit, V_{BB}, V_{CC}, R_E, R_C, and β_{dc} are called the circuit's _____ variables. (7-9)   18. _____

19. In an emitter-biased circuit, V_E, V_C, I_B, I_C, and I_E are called the circuit's _____ variables. (7-9)   19. _____

20. When testing a transistor with an ohmmeter, the reverse to forward resistance measurement ratio for the emitter and collector diodes should be more than _____:1 for silicon. (7-10)   20. _____

III. ILLUSTRATIVE AND PRACTICE PROBLEMS

Illustrative Problem 1. For the given circuit and its transistor's collector curves, do the following:
a. Construct the load line with the circuit's Q point on the transistor's collector curves.
b. From the coordinates of the Q point, determine I_C, V_{CE}, and β_{dc}.

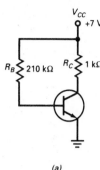
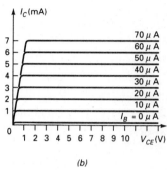

(a)   (b)

Figure 7-1

Steps	Comments
1. Find $I_{C(sat)}$. $$I_{C(sat)} = \frac{7\text{ V}}{1\text{ k}\Omega} = 7\text{ mA}$$	a. Solve for the collector current with the emitter-collector terminals shorted. b. The voltage drop across R_C is V_{CC}. $$I_{C(sat)} = \frac{V_{CC}}{R_C} \quad \text{Ohm's law}$$
2. Find $V_{CE(cut)}$. $$V_{CE(cut)} = 7\text{ V}$$	a. Solve for the collector-emitter voltage with collector-emitter terminals opened. b. There is no collector current. $$V_{CE(cut)} = V_{CC}$$
3. Construct the load line with the circuit's Q point. Figure 7-2 $I_C = 3\text{ mA} \quad V_{CE} = 4\text{ V}$ $$\beta_{dc} = \frac{3\text{ mA}}{30\text{ }\mu\text{A}} = 100$$	a. Draw the load line between $V_{CE(cut)}$ and $I_{C(sat)}$. b. Solve for I_B by applying Ohm's law. $$I_B = \frac{7\text{ V} - 0.7\text{ V}}{210\text{ k}\Omega} = 30\text{ }\mu\text{A}$$ c. The Q point is located where the base current of 30 μA intercepts the load line. d. Locate and show the coordinates of the Q point on the collector curves. e. Solve for β_{dc}: $$\beta_{dc} = \frac{I_C}{I_B}$$

Practice Problem 1. For the given circuit and the transistor's collector curves, do the following:
a. Construct the load line with the circuit's Q point on the transistor's collector curves.
b. From the coordinates of the Q point, determine I_C, V_{CE}, and β_{dc}.

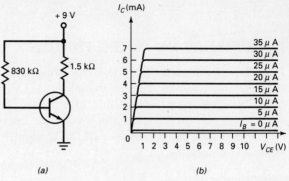

Figure 7-3

Answers:
$I_{C(sat)} = 6$ mA $I_B = 10$ μA $V_{CE} = 6$ V
$V_{CE(cut)} = 9$ V $I_C = 2$ mA $\beta_{dc} = 200$

Illustrative Problem 2. The transistor's dc beta is changed from 100 to 300 in the given circuit. For each β_{dc}, find the following:
a. I_B b. $I_{C(sat)}$ c. I_C d. V_{CE}
e. What region is the transistor operating in?

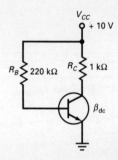

Figure 7-4

Steps	Comments
1. Find I_B. $I_B = \dfrac{10\text{ V} - 0.7\text{ V}}{220\text{ k}\Omega} = 42.3$ μA	a. Solve for I_B by applying Ohm's law. b. In this circuit the base current is independent of β_{dc}.
2. Find $I_{C(sat)}$. $I_{C(sat)} = \dfrac{10\text{ V}}{1\text{ k}\Omega} = 10$ mA	a. Solve for the collector current with the collector-emitter shorted. b. The voltage drop across R_C is V_{CC}. $I_{C(sat)} = \dfrac{V_{CC}}{R_C}$ Ohm's law
3. Find I_C and V_{CE} for $\beta_{dc} = 100$. $I_C = 100(42.3$ μA$) = 4.23$ mA $0 < I_C < 10$ mA active region $V_{CE} = 10\text{ V} - (4.23\text{ mA})(1\text{ k}\Omega) = 5.77$ V	a. Solve for I_C. $I_C = \beta_{dc} I_B$ b. Check to see if $I_C < I_{C(sat)}$. c. If $0 < I_C < I_{C(sat)}$, then the transistor is operating in the active region. d. Solve for V_{CE}. $V_{CE} = V_{CC} - I_C R_C$

Steps	Comments
4. Find I_C and V_{CE} for $\beta_{dc} = 300$. $I_C = 300(42.3\ \mu A) = 12.7$ mA $I_C > 10$ mA saturation region I_C cannot be greater than 10 mA; therefore, $I_C \approx 10$ mA. $V_{CE} = 10\ V - 10\ mA(1\ k\Omega) = 0\ V$	**a.** Solve for I_C. $I_C = \beta_{dc} I_B$ **b.** Check to see if $I_C < I_{C(sat)}$. **c.** If $I_C \geq I_{C(sat)}$, then the transistor is operating in the saturation region and its value is $\approx I_{C(sat)}$ and V_{CE} is ≈ 0 V.

Practice Problem 2. The transistor's dc beta is changed from 100 to 400 in the given circuit. For each β_{dc}, find the following:
a. I_B **b.** $I_{C(sat)}$ **c.** I_C **d.** V_{CE}
e. What region is the transistor operating in?

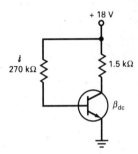

Figure 7-5

Answers:
For $\beta_{dc} = 100$,
a. $I_B = 64.1\ \mu A$ **b.** $I_{C(sat)} = 12$ mA   **c.** $I_C = 6.41$ mA   **d.** $V_{CE} = 8.39$ V   **e.** Active region
For $\beta_{dc} = 400$,
a. $I_B = 64.1\ \mu A$ **b.** $I_{C(sat)} = 12$ mA   **c.** $I_C \approx 12$ mA   **d.** $V_{CE} \approx 0$ V   **e.** Saturation region

Illustrative Problem 3. For the given circuit, whose transistor has a dc beta that is greater than 20, find the following:
a. V_E **b.** $I_{C(sat)}$ **c.** $V_{CE(cut)}$
d. I_E **e.** I_C **f.** V_{CE}

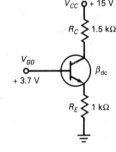

Figure 7-6

Steps	Comments
1. Find V_E. $V_E = 3.7\ V - 0.7\ V = 3\ V$	The transistor's emitter diode is forward-biased. For silicon, there is a 0.7 V drop across it. $V_E = V_{BB} - 0.7\ V$
2. Find $I_{C(sat)}$. $I_{C(sat)} = \dfrac{15\ V - 3\ V}{1.5\ k\Omega} = 8$ mA	**a.** The voltage drop across R_C is $V_{CC} - V_E$, when the collector-emitter terminals are shorted. **b.** Solve for $I_{C(sat)}$ by applying Ohm's law.

Steps	Comments
3. Find $V_{CE(cut)}$. $V_{CE(cut)} = 15\text{ V} - 3\text{ V} = 12\text{ V}$	**a.** Solve for the collector-emitter voltage with the collector-emitter terminals opened. **b.** Since there is no collector current and the emitter diode is forward-biased, $V_{CE(cut)}$ is the voltage difference between V_{CC} and V_E.
4. Find I_E. $I_E = \dfrac{3\text{ V}}{1\text{ k}\Omega} = 3\text{ mA}$	Solve for I_E by applying Ohm's law. $I_E = \dfrac{V_E}{R_E}$
5. Find I_C. $I_C = 3\text{ mA}$	For a $\beta_{dc} > 20$, $I_C \approx I_E$.
6. Find V_C. $V_C = 15\text{ V} - 3\text{ mA}(1.5\text{ k}\Omega) = 10.5\text{ V}$	The voltage at the collector is equal to V_{CC} minus the voltage drop across R_C.
7. Find V_{CE}. $V_{CE} = 10.5\text{ V} - 3\text{ V} = 7.5\text{ V}$	$V_{CE} = V_C - V_E$

Practice Problem 3. For the given circuit, whose transistor has a dc beta that is greater than 20, find the following:

a. V_E b. $I_{C(sat)}$ c. $V_{CE(cut)}$
d. I_E e. I_C f. V_{CE}

Answers: a. $V_E = 3.3\text{ V}$ b. $I_{C(sat)} = 3.22\text{ mA}$ c. $V_{CE(cut)} = 8.7\text{ V}$
d. $I_E = 1\text{ mA}$ e. $I_C = 1\text{ mA}$ f. $V_{CE} = 6\text{ V}$

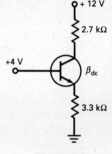

Figure 7-7

IV. PROBLEMS

Sec. 7-1 Variations in Current Gain

7-1. In the circuit of Fig. 7-8, the 2N3904 transistor is operating at its minimum current gain (curves included). If the transistor's junction temperature is 25° C and the voltmeter reading is 10 V, find the following: **a.** β_{dc} **b.** I_B **c.** R_B

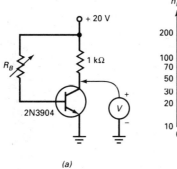
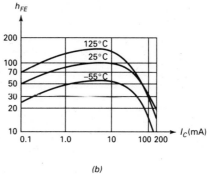

(a) (b)

Figure 7-8

7-2. With respect to Prob. 7-1, if the transistor's junction temperature is increased to 125° C, what value must R_B be adjusted to so that the voltmeter reading is held at 10 V?

Sec. 7-2 The Load Line

7-3. For the circuit of Fig. 7-9, find the following:
  a. $I_{C(sat)}$
  b. $V_{CE(cut)}$

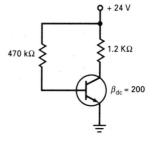

Figure 7-9

7-4. Draw the load lines on the transistor's collector curves for each circuit condition that is shown in Fig. 7-10. ($R_C = 1.2$ kΩ, $R_C = 2.4$ kΩ)

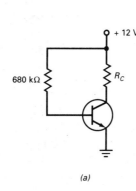
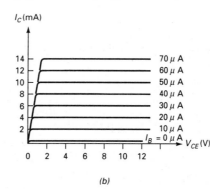

(a) (b)

Figure 7-10

7-5. With respect to Prob. 7-4, select either **increases**, **decreases**, or **stays the same** as the correct answer to the following questions:
  a. What would happen to the cutoff voltage if only R_B is increased?
  b. What would happen to the saturation current if only R_B is increased?
  c. What would happen to the cutoff voltage if only R_C is increased?

d. What would happen to the saturation current if only R_C is increased?
   e. What would happen to the cutoff voltage if only V_{CC} is increased?
   f. What would happen to the saturation current if only V_{CC} is increased?
7-6. With respect to Prob. 7-4, answer the following questions:
   a. What circuit value must be varied in order to change the slope of the load line?
   b. What circuit value must be varied in order to move the load line without changing its slope.
7-7. For the circuit and the transistor's collector curves in Fig. 7-11, find the value of V_{CC} and R_C from the circuit's load line.

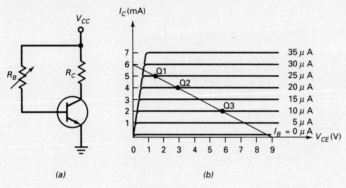

Figure 7-11

Sec. 7-3 The Operating Point

7-8. For the circuit of Fig. 7-12, do the following:
   a. Complete Table 7-1 for the given circuit conditions. Select either *UP* or *DOWN* to indicate the direction the Q point moved along the load line for parts b and c.
   b. In what direction did the Q point move when only R_B was increased?
   c. In what direction did the Q point move when only β_{dc} was increased?

R_B, kΩ	R_C, kΩ	β_{dc}	I_B, μA	I_C, mA	$I_{C(sat)}$, mA	V_{CE}, V
220	1	100				
330	1	100				
220	1	150				
220	1.8	100				

Table 7-1

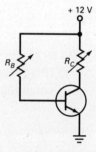

Figure 7-12

7-9. For the circuit and collector curves of Fig. 7-11, do the following:
   a. Complete Table 7-2 for the different Q points that are shown on the circuit's load line.
   b. What circuit value must be changed so that the Q point can move to a different location on the load line for the same set of curves?
   c. Is the Q point coordinate and analytical value of V_{CE} the same?

Q Points	Q-Point Coordinates			Circuit Values		Analytically
	I_B, μA	I_C, mA	V_{CE}, V	R_C, kΩ	R_B, kΩ	$V_{CE} = V_{CC} - I_C R_C$, V
Q_1						
Q_2						
Q_3						

Table 7-2

Sec. 7-5 Recognizing Saturation

7-10. Complete Table 7-3 for the given circuit conditions in Fig. 7-13.

			Meter Readings		
$β_{dc}$	R_B, kΩ	R_C, kΩ	V_1, V	V_2, V	Region of Operation
100	180	1			
100	47	1			
100	180	10			
400	180	1			

Table 7-3

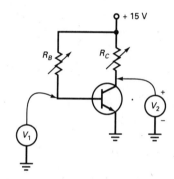

Figure 7-13

7-11. The voltmeter readings for the circuit of Fig. 7-14 suggest that the transistor is in saturation. Indicate whether an increase or decrease in the values of R_B, R_C, and $β_{dc}$ would tend to cause the transistor to come out of saturation.

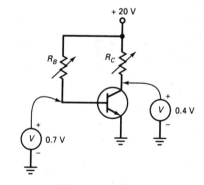

Figure 7-14

Sec. 7-6 The Transistor Switch

7-12. For the circuit of Fig. 7-15, what is the voltmeter reading when the switch is opened? What is the voltmeter reading when the switch is closed?

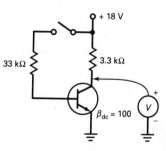

Figure 7-15

Sec. 7-7 Emitter Bias

7-13. For the circuit of Fig. 7-16, find the following:
 a. V_E b. I_E
 c. I_C d. V_C
 e. V_{CE} f. P_D

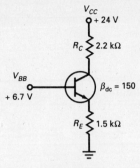

Figure 7-16

7-14. From the voltage measurements for the circuit in Fig. 7-17, find the following:
 a. V_{BB}
 b. V_{CC}

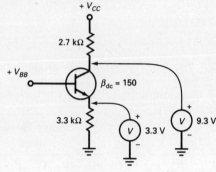

Figure 7-17

7-15. For the circuit of Fig. 7-18, do the following:
 a. Complete Table 7-4, for the given circuit conditions.

V_{CC}, V	R_E, kΩ	R_C, kΩ	$β_{dc}$	I_C, mA	V_{CE}, V
12	1	1.2	100		
12	1	1.2	400		
12	1	1.8	100		
12	1.5	1.2	100		
18	1	1.2	100		

Table 7-4

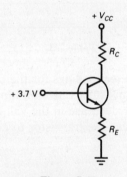

Figure 7-18

Select either *increased*, *decreased*, or *stayed the same* as the correct answer to the following questions.
 b. What happened to I_C when only $β_{dc}$ was increased?
 c. What happened to V_{CE} when only $β_{dc}$ was increased?
 d. What happened to I_C when only R_C was increased?
 e. What happened to V_{CE} when only R_C was increased?
 f. What happened to I_C when only R_E was increased?
 g. What happened to V_{CE} when only R_E was increased?

Sec. 7-8 LED Drivers

7-16. For the circuit of Fig. 7-19, how much current is flowing through the LED when the switch is opened? How much current is flowing through the LED when the switch is closed?

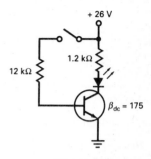

Figure 7-19

7-17. For the circuit of Fig. 7-20, complete Table 7-5 for the given circuit conditions.

V_{BB}, V	R_E, Ω	Diode Current, mA
0	330	
4	330	
4	1 k	

Table 7-5

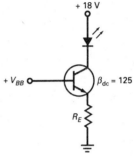

Figure 7-20

Sec. 7-9 The Effects of Small Changes

7-18. For the circuits of Fig. 7-21, complete Table 7-6. [Fill in the table with the letters U (up), D (down), and N (no change) for the circuit's responses to a 10 percent decrease in circuit values.]

Circuit Values, 10% decrease	Circuit Response			
	Base Bias		Emitter Bias	
	I_C, mA	V_{CE}, V	I_C, mA	V_{CE}, V
β_{dc}				
V_{CC}				
V_{BB}				
R_C				
R_B			XXXXX	XXXXXXX
R_E	XXXXX	XXXXXXX		

Table 7-6

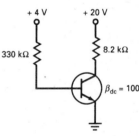

(a)

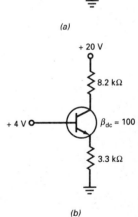

(b)

Figure 7-21

7-19. For the circuit of Fig. 7-22, fill in the readings of the voltmeters in the Table 7-7 for the given circuit conditions.

Trouble	Voltmeter Readings			
	V_1, V	V_2, V	V_3, V	V_4, V
Circuit OK				
R_C shorted				
R_C opened				
Base-emitter diode opened				
Base-collector diode opened				
All transistor terminals opened				
All transistor terminals shorted				
Base supply voltage off				
Collector supply voltage off				

Table 7-7

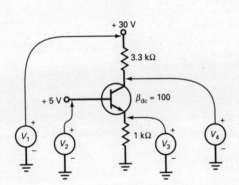

Figure 7-22

CHAPTER 8
TRANSISTOR BIASING

Study Chap. 8 in *Electronic Principles*.

I. TRUE / FALSE

Answer true (T) or false (F) to each of the following statements.

Answer

1. The voltage applied to the base of a transistor from a fixed power supply can be reduced by inserting a voltage divider network. **(8-1)** 1. ___

2. Whenever two unconnected nodes are of the same potential, you can connect a wire between them without changing the operation of the circuit. **(8-1)** 2. ___

3. An assumption that can be made when doing voltage-divider bias analysis is that the base current has no effect on the voltage divider. **(8-2)** 3. ___

4. Once the base voltage of the voltage-divider-biased (VDB) circuit is found, the emitter voltage can easily be determined. **(8-2)** 4. ___

5. When a VDB circuit is properly designed, variations in the transistor's current gain will cause the Q point to move. **(8-2)** 5. ___

6. In a VDB circuit, if the bottom resistor R_2 of the voltage divider is increased, the base voltage will decrease. **(8-2)** 6. ___

7. In a VDB circuit, if the emitter resistor is increased, the emitter voltage will stay the same. **(8-3)** 7. ___

8. In a VDB circuit, the collector supply voltage controls the transistor's cutoff voltage and saturation current. **(8-3)** 8. ___

9. In a VDB circuit, the collector resistor controls the transistor's saturation current. **(8-3)** 9. ___

10. In a two-supply emitter-biased (TSEB) circuit with an *npn* transistor, the negative supply reverse-biases the emitter diode. **(8-4)** 10. ___

11. If the collector resistance is decreased in a TSEB circuit, the collector current will increase. **(8-4)** 11. ___

12. In a TSEB circuit with an *npn* transistor, the negative supply forward-biases the emitter diode. **(8-4)** 12. ___

13. If the transistor's current gain of a TSEB circuit increases, the collector current will increase. **(8-4)** 13. ___

14. If the positive supply of a TSEB circuit with an *npn* transistor is increased, the collector current will increase. **(8-4)** 14. ___

15. If you have a circuit with *npn* transistors, you can often use the same circuit with a negative power supply and *pnp* transistors. (8-5)   15. ___
16. The collector current is greater than the emitter current in a *pnp* transistor. (8-5)   16. ___
17. The collector and emitter diodes of a *pnp* transistor are in the same direction as the diodes of an *npn* transistor. (8-5)   17. ___
18. If a *pnp* transistor is properly biased, the collector voltage will be greater than the emitter voltage. (8-5)   18. ___
19. Negative feedback exists when a decrease of an output quantity produces increases in an input quantity. (8-6)   19. ___
20. In an emitter-feedback bias circuit, an increase in emitter current will produce a decrease in base current. (8-6)   20. ___

II. COMPLETION

Complete each of the following statements.

Answer

1. A VDB circuit can be derived from the ____-biased circuit. (8-1)   1. ___
2. The number of power supplies a VDB circuit has is ____. (8-1)   2. ___
3. With a VDB, a fixed base voltage means a fixed ____ voltage. (8-2)   3. ___
4. The usual design rule for a VDB circuit is that the base current should be at least ____ times smaller than the current through the voltage divider. (8-2)   4. ___
5. When the base current is less than 1 percent of the voltage divider's current, the variation in base voltage will also be less than 1 percent and the voltage divider will be considered ____. (8-2)   5. ___
6. In a VDB circuit, if the top resistor R_1 of the voltage divider is increased, the base voltage will ____. (8-2)   6. ___
7. In a VDB circuit, if the base voltage is decreased, the emitter voltage will ____. (8-2)   7. ___
8. In a VDB circuit, if the emitter resistor is decreased, the collector current will ____. (8-3)   8. ___
9. In a VDB circuit, if the emitter voltage is increased, the collector current will ____. (8-3)   9. ___
10. In a TSEB circuit with an *npn* transistor, the positive supply ____-biases the collector diode. (8-4)   10. ___
11. The TSEB circuit is derived from the ____ bias circuit. (8-4)   11. ___
12. If the TSEB circuit is properly designed, the value of the base voltage is approximately ____ V. (8-4)   12. ___
13. If the negative supply of a TSEB circuit with an *npn* transistor is decreased, the collector current will ____. (8-4)   13. ___
14. If the TSEB circuit is properly designed for an *npn* silicon transistor, the value of the emitter voltage is approximately ____ V. (8-4)   14. ___
15. If a *pnp* transistor is properly biased, the algebraic sign for V_{CE} is always ____. (8-5)   15. ___
16. If a *pnp* transistor is properly biased, the algebraic sign for V_{BE} is always ____. (8-5)   16. ___
17. For the emitter diode of a *pnp* transistor to be forward-biased, the voltage at the base must be ____ than the voltage at the emitter. (8-5)   17. ___
18. If a *pnp* transistor is properly biased, the collector voltage will be ____ than the emitter voltage. (8-5)   18. ___
19. The phenomenon that is characterized by how changes of an output quantity causes changes of an input quantity is called ____. (8-6)   19. ___
20. The most popular biasing circuit configuration because of its Q-point stability is the ____-____ bias circuit. (8-6)   20. ___

III. HIGHLIGHTS OF CIRCUITS

First, to find I_C, apply Kirchoff's voltage law (KVL) around the closed loop that includes voltage V_{BE}. Then assume that $I_C \approx I_E$, since β_{dc} is almost always greater than 20. Finally, I_C cannot be greater than $I_{C(sat)}$. The theoretical value of $I_{C(sat)}$ can be determined by setting V_{CE} to zero. When I_C is calculated and turns out to be greater than $I_{C(sat)}$, this indicates that the transistor is in saturation. In saturation, I_C is limited to the value of $I_{C(sat)}$. If, however, the transistor is forced into hard saturation, the β_{dc} can be drastically reduced. When β_{dc} is reduced enough, I_C can decrease to a value that is less than the theoretical $I_{C(sat)}$, providing that the circuit has an emitter resistor.

1. *Emitter-Feedback Bias*

$$I_C = \frac{V_{CC} - 0.7 \text{ V}}{R_E + R_B/\beta_{dc}}$$

I_C must be less than $I_{C(sat)}$.

$$I_{C(sat)} = \frac{V_{CC}}{R_C + R_E}$$

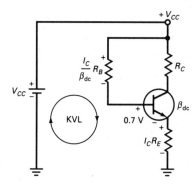

Figure 8-1

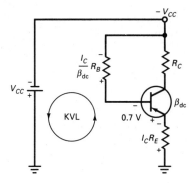

Figure 8-2

2. *Collector-Feedback Bias*

$$I_C = \frac{V_{CC} - 0.7 \text{ V}}{R_C + R_B/\beta_{dc}}$$

I_C must be less than $I_{C(sat)}$.

$$I_{C(sat)} = \frac{V_{CC}}{R_C}$$

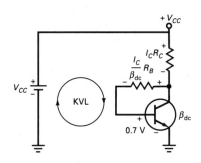

Figure 8-3

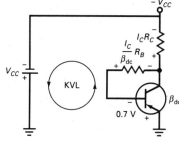

Figure 8-4

3. *Collector- and Emitter-Feedback Bias*

$$I_C = \frac{V_{CC} - 0.7 \text{ V}}{R_C + R_E + R_B/\beta_{dc}}$$

I_C must be less than $I_{C(sat)}$.

$$I_{C(sat)} = \frac{V_{CC}}{R_C + R_E}$$

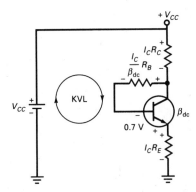

Figure 8-5

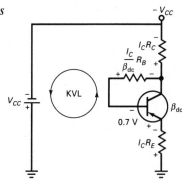

Figure 8-6

4. Two-Supply Emitter Bias

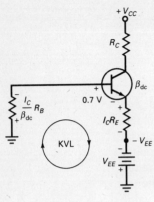

Figure 8-7

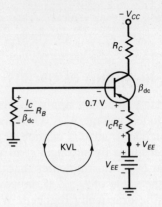

Figure 8-8

$$I_C = \frac{V_{EE} - 0.7 \text{ V}}{R_E + R_B/\beta_{dc}}$$

If $\beta_{dc}R_E$ is greater than $20R_B$, then R_E swamps out R_B/β_{dc}.

$$I_C \approx \frac{V_{EE} - 0.7 \text{ V}}{R_E}$$

I_C must be less than $I_{C(sat)}$.

$$I_{C(sat)} = \frac{V_{EE} + V_{CC}}{R_C + R_E}$$

5. Voltage-Divider Bias

$R_{TH} = R_1 \parallel R_2$

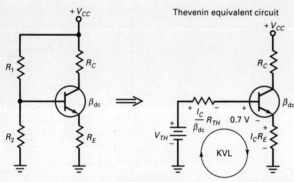

Figure 8-9

$$V_{TH} = \frac{R_2}{R_1 + R_2} V_{CC}$$

$$I_C = \frac{V_{TH} - 0.7 \text{ V}}{R_E + R_B/\beta_{dc}}$$

If $\beta_{dc}R_E$ is greater than $20R_{TH}$, then R_E swamps out R_B/β_{dc}.

$$I_C \approx \frac{V_{TH} - 0.7 \text{ V}}{R_E}$$

I_C must be less than $I_{C(sat)}$.

$$I_{C(sat)} = \frac{V_{CC}}{R_C + R_E}$$

$R_{TH} = R_1 \parallel R_2$

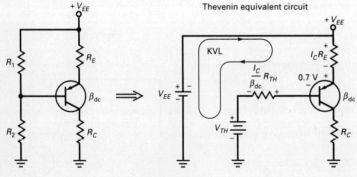

Figure 8-10

$$V_{TH} = \frac{R_2}{R_1 + R_2} V_{EE}$$

$$I_C = \frac{V_{EE} - V_{TH} - 0.7 \text{ V}}{R_E + R_B/\beta_{dc}}$$

If $\beta_{dc}R_E$ is greater than $20R_{TH}$, then R_E swamps out R_B/β_{dc}.

$$I_C \approx \frac{V_{EE} - V_{TH} - 0.7 \text{ V}}{R_E}$$

I_C must be less than $I_{C(sat)}$.

$$I_{C(sat)} = \frac{V_{EE}}{R_C + R_E}$$

III. ILLUSTRATIVE AND PRACTICE PROBLEMS

Illustrative Problem 1. Find the collector current for the given VDB circuit. [Assume that the transistor is not in saturation. I_C must be less than $I_{C(sat)}$.]

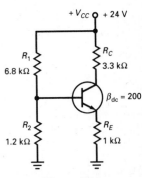

Figure 8-11

Steps	Comments
1. Find V_{TH} and R_{TH}. Figure 8-12 $V_{TH} = \dfrac{1.2 \text{ k}\Omega}{6.8 \text{ k}\Omega + 1.2 \text{ k}\Omega} \; 24 \text{ V} = 3.6 \text{ V}$ $R_{TH} = 1.2 \text{ k}\Omega \; // \; 6.8 \text{ k}\Omega = 1.02 \text{ k}\Omega$	Thevenize the voltage divider network with respect to the base. $V_{TH} = \dfrac{R_2}{R_1 + R_2} V_{CC}$ $R_{TH} = R_1 \; // \; R_2$
2. Find I_C. Figure 8-13 $I_C = \dfrac{3.6 \text{ V} - 0.7 \text{ V}}{1 \text{ k}\Omega + 1.02 \text{ k}\Omega / 200} = 2.89 \text{ mA}$ $I_{C(sat)} = \dfrac{24 \text{ V}}{3.3 \Omega + 1 \text{ k}\Omega} = 5.58 \text{ mA}$	a. Draw the Thevenin equivalent circuit. b. The VDB circuit becomes an equivalent base bias circuit. c. For a $\beta_{dc} > 20$, $I_C \approx I_E$. d. Apply KVL around the closed loop that includes V_{BE} and solve for I_C. $I_C = \dfrac{V_{TH} - 0.7 \text{ V}}{R_E + R_{TH}/\beta_{dc}}$ e. Check to see that $I_C < I_{C(sat)}$.
3. Find I_C (approximate). $\beta_{dc} R_E = 200 \text{ k}\Omega \qquad 20 R_{TH} = 20.4 \text{ k}\Omega$ $I_C = \dfrac{3.6 \text{ V} - 0.7 \text{ V}}{1 \text{ k}\Omega} = 2.9 \text{ mA}$	If $\beta_{dc} R_E > 20 R_{TH}$, then R_E swamps out R_{TH}/β_{dc} and I_C becomes $I_C = \dfrac{(V_{TH} - 0.7 \text{ V})}{R_E}$ (becomes the voltage drop across R_E) and $V_{TH} \approx V_B$.

Practice Problem 1. Find the collector current for the given VDB circuit. [Assume that the transistor is not in saturation. I_C must be less than $I_{C(sat)}$.]

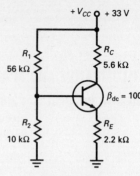

Figure 8-14

Answers: $I_C = 1.88$ mA $I_C = 1.95$ mA   (approximate)

Illustrative Problem 2. For the given VDB circuit, find the following. [Assume that the transistor is not in saturation. This means that for a *pnp* transistor, $|V_{CE}|$ should be greater than zero and I_C must be less than $I_{C(sat)}$.]
a. I_C
b. V_{CE}

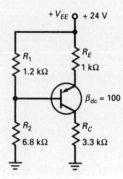

Figure 8-15

Steps	Comments
1. Find V_{TH} and R_{TH}. $$V_{TH} = \frac{6.8 \text{ k}\Omega}{6.8 \text{ k}\Omega + 1.2 \text{ k}\Omega} 24\text{V} = 20.4 \text{ V}$$ $$R_{TH} = 1.2 \text{ k}\Omega \; // \; 6.8 \text{ k}\Omega = 1.02 \text{ k}\Omega$$	Thevenize the voltage divider network. $$V_{TH} = \frac{R_2}{R_1 + R_2} V_{CC}$$ $$R_{TH} = R_1 \; // \; R_2$$
2. Find I_C. Figure 8-16 $$I_C = \frac{24 \text{ V} - 20.4 \text{ V} - 0.7 \text{ V}}{1 \text{ k}\Omega + 1.02 \text{ k}\Omega/100} = 2.87 \text{ mA}$$ $$I_{C(sat)} = \frac{24 \text{ V}}{3.3 \text{ k}\Omega + 1 \text{ k}\Omega} = 5.58 \text{ mA}$$	a. Draw the Thevenin equivalent circuit. b. The VDB circuit becomes an equivalent base bias circuit. c. For a $\beta_{dc} > 20$, $I_C \approx I_E$. d. Apply KVL around the closed loop that includes V_{BE} and solve for I_C. $$I_C = \frac{(V_{EE} - V_{TH} - 0.7 \text{ V})}{R_E + R_{TH}/\beta_{dc}}$$ e. Check to see that $I_C < I_{C(sat)}$. $$I_{C(sat)} = \frac{V_{EE}}{R_C + R_E}$$

Steps	Comments
3. Find I_C (approximate). $\beta_{dc}R_E = 100 \text{ k}\Omega \quad 20R_{TH} = 20.4 \text{ k}\Omega$ $I_C = \dfrac{24 \text{ V} - 20.4 \text{ V} - 0.7 \text{ V}}{1 \text{ k}\Omega} = 2.9 \text{ mA}$	If $\beta_{dc}R_E > 20R_{TH}$, then R_E swamps out R_{TH}/β_{dc} and I_C becomes $I_C = \dfrac{(V_{EE} - V_{TH} - 0.7 \text{ V})}{R_E}$ (becomes the voltage drop across R_E) and $V_{TH} \approx V_B$.
4. Find V_E. $V_E = 24 \text{ V} - (2.9 \text{ mA})(1 \text{ k}\Omega) = 21.1 \text{ V}$	Use the approximate value of I_C. $V_E = V_{EE} - I_C R_E$
5. Find V_C. $V_C = (2.9 \text{ mA})(3.3 \text{ k}\Omega) = 9.57 \text{ V}$	Use the approximate value of I_C. $V_C = I_C R_C$
6. Find V_{CE}. $V_{CE} = 9.57 \text{ V} - 21.1 \text{ V} = -11.53 \text{ V}$	$V_{CE} = V_C - V_E$

Practice Problem 2. For the given VDB circuit, find the following. [Assume that the transistor is not in saturation. This means that for a *pnp* transistor, $|V_{CE}|$ should be greater than zero and I_C must be less than $I_{C(\text{sat})}$.]
  a. I_C (If $\beta_{dc}R_E > 20R_{TH}$, approximate I_C.)
  b. V_{CE}

Answers: $I_C = 1.68 \text{ mA} \quad V_{CE} = -18.4 \text{ V}$

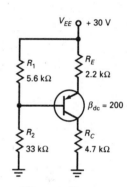

Figure 8-17

Illustrative Problem 3. For the given TSEB circuit, find the following. [Assume that the transistor is not in saturation. This means that for an *npn* transistor, V_{CE} should be greater than zero and I_C must be less than $I_{C(\text{sat})}$.]
  a. I_C
  b. V_{CE}

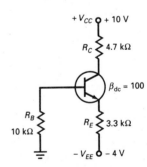

Figure 8-18

Steps	Comments
1. Find I_C. Figure 8-19 $$I_C = \frac{4\text{ V} - 0.7\text{ V}}{3.3\text{ k}\Omega + 10\text{ k}\Omega/100} = 0.97\text{ mA}$$ $$I_{C(sat)} = \frac{10\text{ V} + 4\text{ V}}{3.3\text{ k}\Omega + 4.7\text{ k}\Omega} = 1.75\text{ mA}$$	a. For a $\beta_{dc} > 20$, $I_C \approx I_E$. b. Apply KVL around the closed loop that includes V_{BE} and solve for I_C. $$I_C = \frac{V_{EE} - 0.7\text{ V}}{R_E + R_B/\beta_{dc}}$$ c. Check to see that $I_C < I_{C(sat)}$. $$I_{C(sat)} = \frac{V_{CC} + V_{EE}}{R_C + R_E}$$
2. Find I_C (approximate). $\beta_{dc}R_E = 330\text{ k}\Omega \qquad 20R_B = 200\text{ k}\Omega$ $$I_C = \frac{4\text{ V} - 0.7}{3.3\text{ k}\Omega} = 1\text{ mA}$$	If $\beta_{dc}R_E > 20R_B$, then R_E swamps out R_{TH}/β_{dc} and I_C becomes $$I_C = \frac{(V_{EE} - 0.7\text{ V})}{R_E}$$ (becomes the voltage drop across the R_E) and $V_B \approx 0\text{ V}$.
3. Find V_E. $V_E = -4\text{ V} + (1\text{ mA})(3.3\text{ k}\Omega) = -0.7\text{ V}$	Use the approximate value of I_C. $V_E = -V_{EE} + I_C R_E$
4. Find V_C. $V_C = 10\text{ V} - (1\text{ mA})(4.7\text{ k}\Omega) = 5.3\text{ V}$	Use the approximate value of I_C. $V_C = V_{CC} - I_C R_C$
5. Find V_{CE}. $V_{CE} = 5.3\text{ V} - (-0.7\text{ V}) = 6\text{ V}$	$V_{CE} = V_C - V_E$

Practice Problem 3. For the given TSEB circuit, find the following. [Assume that the transistor is not in saturation. This means that for an *npn* transistor, V_{CE} must be greater than zero and $I_C < I_{C(sat)}$.]
  a. I_C (If $\beta_{dc}R_E > 20R_B$, approximate I_C.)
  b. V_{CE}

Answers: $I_C = 5.85\text{ mA} \qquad V_{CE} = 10.2\text{ V}$

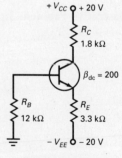

Figure 8-20

V. PROBLEMS

Sec. 8-2 VDB Analysis

8-1. For the circuit of Fig. 8-21, do the following:
  a. Complete Table 8-1 for the given circuit conditions. (For this problem do not find the approximate value of I_C, even if $\beta_{dc}R_E > 20R_{TH}$.)

R_E, kΩ	R_C, kΩ	β_{dc}	I_C, mA	V_{CE}, V
3.3	10	100		
6.8	10	100		
3.3	18	100		
3.3	10	200		

Table 8-1

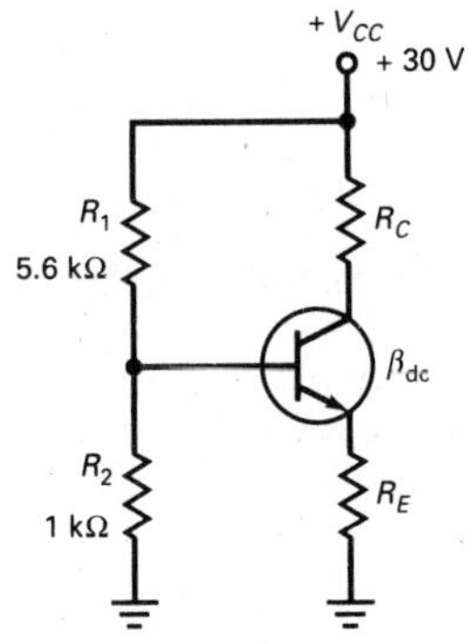

Figure 8-21

Select either *increased*, *decreased*, or *stayed the same* as the correct answer to the following questions.
  b. What happened to I_C when only R_E was increased?
  c. What happened to V_{CE} when only R_E was increased?
  d. What happened to I_C when only R_C was increased?
  e. What happened to V_{CE} when only R_C was increased?
  f. What happened to I_C when only β_{dc} was increased?
  g. What happened to V_{CE} when only β_{dc} was increased?

8-2. For the circuits of Fig. 8-22, determine what circuit yields a smaller error when the approximate value of I_C is calculated.

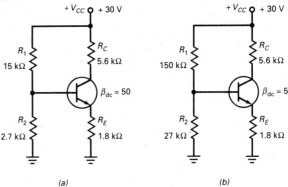

Figure 8-22

8-3. From the voltmeter readings for the circuit of Fig. 8-23, find the following:
  a. V_{CE}
  b. I_C
  c. R_C

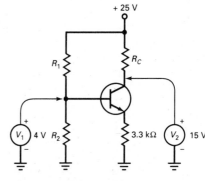

Figure 8-23

8-4. Determine what circuit or circuits of Fig. 8-24 are in saturation.

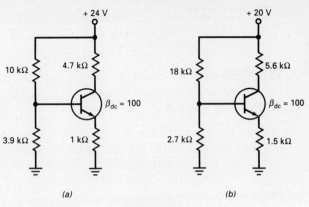

Figure 8-24

Sec. 8-4 Two-Supply Emitter Bias

8-5. For the circuit of Fig. 8-25, do the following:
  a. Complete Table 8-2 for the given circuit conditions. (For this problem do not find the approximate value of I_C, even if $\beta_{dc}R_E > 20R_B$.)

R_E, kΩ	R_C, kΩ	β_{dc}	I_C, mA	V_{CE}, V
5.6	3.3	100		
10	3.3	100		
5.6	4.7	100		
5.6	3.3	200		

Table 8-2

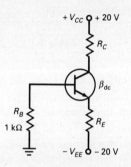

Figure 8-25

(Select either *increased*, *decreased*, or *stayed the same* as the correct answer to the following questions).
  b. What happened to I_C when only R_E was increased?
  c. What happened to V_{CE} when only R_E was increased?
  d. What happened to I_C when only R_C was increased?
  e. What happened to V_{CE} when only R_C was increased?
  f. What happened to I_C when only β_{dc} was increased?
  g. What happened to V_{CE} when only β_{dc} was increased?

8-6. For the circuits of Fig. 8-26, determine what circuit yields a smaller error when the approximate value of I_C is calculated.

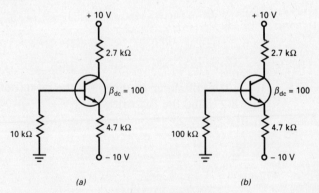

Figure 8-26

8-7. Determine what circuit or circuits of Fig. 8-27 are in saturation.

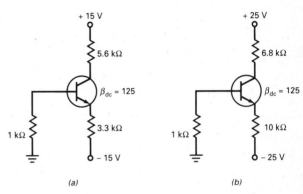

Figure 8-27

8-8. From the voltmeter readings for the circuit of Fig. 8-28, find the following:
a. V_{CE}
b. I_C
c. R_C

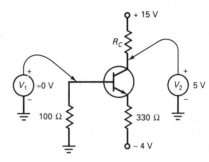

Figure 8-28

8-9. For the circuit of Fig. 8-29, find the following:
a. I_E
b. I_C
c. V_{CE}
d. P_D

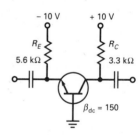

Figure 8-29

Sec. 8-5 PNP Transistors

Note: Find the approximate value of I_C when $\beta_{dc} R_E > 20 R_{TH}$.

8-10. For the circuit of Fig. 8-30, find the following:
a. I_C
b. V_{CE}

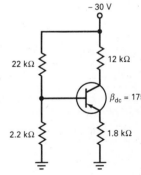

Figure 8-30

8-11. For the circuit of Fig. 8-31, find the following:
  a. I_E
  b. V_{CE}

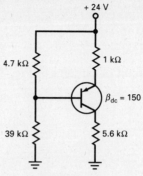

Figure 8-31

8-12. For the circuit of Fig. 8-32, find the following:
  a. I_C
  b. V_{CE}

8-13. For the circuit of Fig. 8-33, find the following:
  a. I_E
  b. V_{CE}

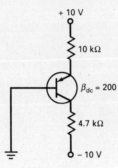

Figure 8-32

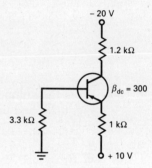

Figure 8-33

Sec. 8-6 Other Types of Biasing

8-14. For the circuit of Fig. 8-34, find the following:
  a. I_C
  b. V_{CE}
  c. P_D

8-15. From the voltage measurements for the circuit of Fig 8-35, find the following:
  a. I_E
  b. β_{dc}
  c. R_E

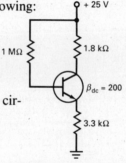

Figure 8-34

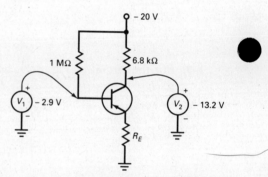

Figure 8-35

8-16. For the circuit of Fig. 8-36, find the following:
  a. I_C
  b. V_{CE}
  c. P_D

8-17. From the voltage measurement for the circuit of Fig. 8-37, find the following:
  a. I_E
  b. β_{dc}

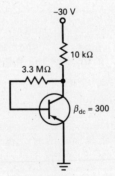

Figure 8-36

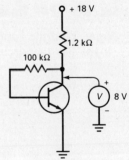

Figure 8-37

8-18. For the circuit of Fig. 8-38, find the following:
  a. I_C
  b. V_{CE}
  c. P_D

8-19. From the voltage measurements for the circuit of Fig. 8-39, find the following:
  a. I_E
  b. β_{dc}
  c. R_E

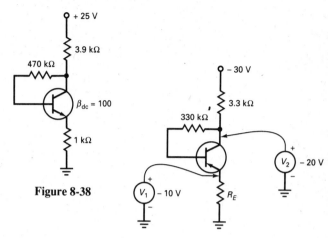

Figure 8-38

Figure 8-39

Sec. 8-7 Troubleshooting

8-20. For the circuit of Fig. 8-40, fill in the readings of the voltmeters in Table 8-3 for the given circuit conditions.

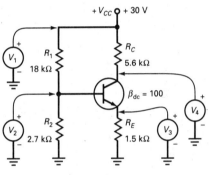

Figure 8-40

Trouble	Voltmeter Readings			
	V_1, V	V_2, V	V_3, V	V_4, V
Circuit OK				
R_1 opened				
R_2 shorted				
Base-emitter diode opened				
All transistor terminals opened				
All transistor terminals shorted				
R_C shorted				
Collector supply voltage off				

Table 8-3

Name _____ Date _____

CHAPTER 9
AC MODELS

Study Chap. 9 in *Electronic Principles*.

I. TRUE / FALSE

Answer true (T) or false (F) to each of the following statements.

Answer

1. In amplifiers, capacitors are only used to couple or transmit ac signals from one circuit to another. **(9-1)**   1. ___
2. A capacitor is shorted at high frequencies and opened at low frequencies. **(9-1)**   2. ___
3. The current through a capacitor increases to the maximum value when the capacitor's reactance goes to zero. **(9-1)**   3. ___
4. At the critical frequency, the current through a capacitor is 50 percent of the maximum value. **(9-1)**   4. ___
5. The high-frequency border for a coupling capacitor is 10 times greater than its critical frequency. **(9-1)**   5. ___
6. At high frequencies, a bypass capacitor shunts practically all of the ac current away from the resistor. **(9-2)**   6. ___
7. The frequency where the reactance of the bypass capacitor equals the Thevenin resistance facing it is called the critical frequency. **(9-2)**   7. ___
8. The simplest way to analyze an amplifier circuit is to split the analysis into two parts by using the superposition theorem. **(9-3)**   8. ___
9. The total voltage at any node or across any resistor of an amplifier is the sum of its ac and dc voltage components. **(9-3)**   9. ___
10. When an ac voltage is coupled onto the base of a biased transistor, the operating point is forced to move up and down about its quiescent point. **(9-4)**   10. ___
11. Small-signal amplifiers are amplifiers that have a peak-to-peak emitter current that is less than 25 percent of the dc emitter current. **(9-4)**   11. ___
12. The stretching and compressing of alternate half-cycles of the emitter current produced by the nonlinear properties of the emitter diode is called distortion. **(9-4)**   12. ___
13. The ac resistance of a diode or transistor is the same as its dc resistance. **(9-5)**   13. ___
14. The ac resistance of an emitter diode equals the ac base-emitter voltage divided by the dc emitter current. **(9-5)**   14. ___
15. The ac current gain of the transistor equals a change of collector current divided by the corresponding change in base current. **(9-6)**   15. ___

Copyright © 1993 by the Glencoe Division of Macmillan/McGraw-Hill School Publishing Company. All rights reserved.

16. The output capacitor of a common emitter (CE) amplifier blocks the dc collector voltage but passes the ac collector voltage to the load. (9-7)   16. ___
17. The input impedance of a CE stage is the combined effect of the biasing resistance and the input impedance of the base. (9-8)   17. ___
18. In a CE stage, $z_{in(base)}$ is smaller than z_{in}. (9-8)   18. ___
19. In analysis and design, h parameters are used to model the transistor based on what is happening at its terminals without regard to what is occurring internally. (9-9)   19. ___

II. COMPLETION

Complete each of the following statements.

Answer

1. If the frequency is increased, the capacitance reactance of a capacitor will ____. (9-1)   1. ___
2. In order to neglect the effects of the reactance of a coupling capacitor, it should be at least ____ times smaller than the total resistance that is in series with it. (9-1)   2. ___
3. The frequency where the reactance of the coupling capacitor equals the total resistance that is in series with it is called the ____ frequency. (9-1)   3. ___
4. When the frequency is greater than the high-frequency border, the coupling capacitor acts like an ac ____. (9-1)   4. ___
5. For dc, the current through a capacitor is ____. (9-1)   5. ___
6. The kind of ground that is produced by a bypass capacitor and only exists at high frequencies is called an ____ ground. (9-2)   6. ___
7. The emitter bypass capacitor creates an ____ ground at the emitter. (9-2)   7. ___
8. In analyzing the dc equivalent of an amplifier, reduce the ac source to zero and ____ all the capacitors. (9-3)   8. ___
9. In analyzing the ac equivalent of an amplifier, reduce all the dc sources to zero and ____ all the capacitors. (9-3)   9. ___
10. In order to keep distortion minimal, the peak-to-peak value of the ac emitter current should be kept less than ____ percent of the dc emitter current. (9-4)   10. ___
11. Using calculus, the ac resistance of the emitter diode can be shown to equal ____ mV divided by the dc current in milliamperes through the emitter diode. (9-5)   11. ___
12. If you change the biasing to get more dc current, the ac resistance of the emitter diode will ____. (9-5)   12. ___
13. The symbol for the ac current gain of the transistor is ____. (9-6)   13. ___
14. Graphically, the slope of the curve for the current gain of the transistor is the ____ current gain. (9-6)   14. ___
15. The output voltage of a CE amplifier is out of phase from the input voltage by ____ degrees. (9-7)   15. ___
16. The ac voltage at the emitter of a CE amplifier is 0 V due to the presence of a ____ capacitor. (9-7)   16. ___
17. The input impedance of a CE transistor looking into the base equals the ac base voltage divided by the ac ____ current. (9-8)   17. ___
18. The ac base-emitter voltage divided by the ac emitter diode resistance equals the ac ____ current. (9-8)   18. ___
19. The symbol for the ac current gain given in the "Small-Signal Characteristics" section of the data sheet is ____. (9-9)   19. ___
20. Quantities like β and r'_e are called ____ parameters. (9-9)   20. ___

82 Copyright © 1993 by the Glencoe Division of Macmillan/McGraw-Hill School Publishing Company. All rights reserved.

III. ILLUSTRATIVE AND PRACTICE PROBLEMS

Illustrative Problem 1. For the CE amplifier, do the following:
    a. Draw the dc equivalent circuit.
    b. Find I_E.
    c. Find V_{CE}.
    d. Find r'_e.

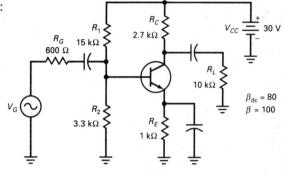

Figure 9-1

Steps	Comments
1. Draw the dc equivalent circuit. Figure 9-2	a. Reduce the ac voltage source to zero. b. Open all capacitors.
2. Find I_E. $V_{TH} = \dfrac{3.3\ \text{k}\Omega}{3.3\ \text{k}\Omega + 15\ \text{k}\Omega}\, 30\ \text{V} = 5.41\ \text{V}$ $V_E = 5.41\ \text{V} - 0.7\ \text{V} = 4.71\ \text{V}$ $I_E = \dfrac{4.71\ \text{V}}{1\ \text{k}\Omega} = 4.71\ \text{mA}$	a. Since $\beta_{dc} > 20$, $I_C \approx I_E$. b. Calculate V_{TH}. $V_{TH} = \dfrac{R_2}{R_1 + R_2}\, V_{CC}$ c. Calculate V_E. Since $\beta_{dc} R_E > 20 R_{TH}$, $V_{TH} \approx V_B$. Therefore, $V_E = V_B - 0.7\ \text{V}$ d. $I_E = \dfrac{V_E}{R_E}$
3. Find V_{CE}. $V_{CE} = 30\ \text{V} - 4.71\ \text{mA}(2.7\ \text{k}\Omega + 1\ \text{k}\Omega) = 12.6\ \text{V}$	$V_{CE} = V_{CC} - I_C(R_C + R_E)$ Ideally, for midpoint biasing $V_{CE} \approx \dfrac{V_{CC}}{2}$
4. Find r'_e. $r'_e = \dfrac{25\ \text{mV}}{4.71\ \text{mA}} = 5.31\ \Omega$	$r'_e = \dfrac{25\ \text{mV}}{I_E}$

Practice Problem 1. For the CE amplifier, do the following:
a. Draw the dc equivalent circuit.
b. Find I_E.
c. Find V_{CE}.
d. Find r'_e.

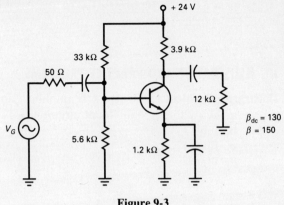

Figure 9-3

Answers: **b.** $I_E = 2.32$ mA **c.** $V_{CE} = 12.2$ V **d.** $r'_e = 10.8$ Ω

Illustrative Problem 2. For the CE amplifier, do the following (given $r'_e = 5.31$ Ω):
a. Draw the ac equivalent circuit.
b. Find $z_{in(base)}$.
c. Find z_{in}.

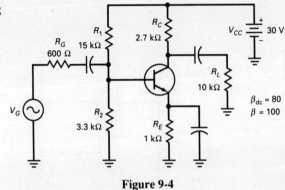

Figure 9-4

Steps	Comments
1. Draw the ac equivalent circuit. ![Figure 9-5] Figure 9-5 ![Figure 9-6] Figure 9-6	a. Reduce the dc voltage source to zero. b. Short all capacitors. c. Combine all the resistors that are in parallel. $R_1 // R_2 = \dfrac{(15 \text{ k}\Omega)(3.3 \text{ k}\Omega)}{(15 \text{ k}\Omega + 3.3 \text{ k}\Omega)} = 2.7 \text{ k}\Omega$ $R_C // R_L = \dfrac{(2.7 \text{ k}\Omega)(10 \text{ k}\Omega)}{(2.7 \text{ k}\Omega + 10 \text{ k}\Omega)} = 2.13 \text{ k}\Omega$ d. Show the ac equivalent circuit with all the resistors that are in parallel combined.

Steps	Comments
2. Find $z_{in(base)}$. $$z_{in(base)} = 100(5.31\ \Omega) = 531\ \Omega$$	Input impedance of the transistor looking into the base: $$z_{in(base)} = \beta r'_e$$
3. Find z_{in}. $$z_{in} = \frac{(2.7\ k\Omega)(0.531\ k\Omega)}{(2.7\ k\Omega + .531\ k\Omega)} = 444\ \Omega$$	Input impedance of the stage: $$z_{in} = R_1 // R_2 // \beta r'_e$$

Practice Problem 2. For the CE amplifier, do the following (given $r'_e = 10.8\ \Omega$).
  a. Draw the ac equivalent circuit.
  b. Find $z_{in(base)}$.
  c. Find z_{in}.

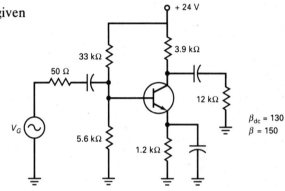

Figure 9-7

Answers: **b.** $z_{in(base)} = 1.62\ k\Omega$ **c.** $z_{in} = 1.21\ k\Omega$

IV. PRACTICAL TECHNIQUES

Objective:

Measure the ac current gain of a transistor. (Remember that $\beta = \Delta I_C / \Delta I_B$ with V_{CE} held constant.)

Steps	Comments
1. Construct the circuit that is shown in Fig. 9-8. **Figure 9-8**	a. R_B (0–1 MΩ decade resistor box). b. R_C (0–10 kΩ decade resistor box). c. Set R_B to 1 MΩ. d. Set R_C so that voltmeter V_2 reads approximately 10 V. e. Record the reading of R_C. f. Record the readings of voltmeters.
2. Determine I_B. $I_{B1} = (20\text{ V} - V_1) \div 1\text{ M}\Omega$	Calculate I_B from the measured values. For this calculation label it I_{B1}.
3. Determine I_C. $I_{C1} = (20\text{ V} - V_2) \div R_C$	Calculate I_C from the measured values. For this calculation label it I_{C1}.
4. Vary I_B and I_C a small amount.	Decrease R_B by 10%. Set R_B to 0.9 MΩ.
5. Adjust R_C.	a. Adjust R_C so that voltmeter V_2 still reads approximately 10 V. b. Record the reading of R_C. Label it R'_C. c. Record the readings of voltmeters. Label them V'_1 and V'_2.
6. Determine I_B. $I_{B2} = (20\text{ V} - V'_1) \div 0.9\text{ M}\Omega$	Calculate I_B from the measured values. For this calculation label it I_{B2}.
7. Determine I_C. $I_{C2} = (20\text{ V} - V'_2) \div R'_C$	Calculate I_C from the measured values. For this calculation label it I_{C2}.
8. Determine β. $\beta = \dfrac{\Delta I_C}{\Delta I_B}$	Calculate β from your measurements. $\Delta I_C = I_{C2} - I_{C1}$ $\Delta I_B = I_{B2} - I_{B1}$

Example 1. From the data of circuit measurements for the circuit of Fig. 9-8, determine the ac current gain of the transistor. Data:

1. $R_B = 1$ MΩ, $R_C = 2.9$ kΩ, $V_1 = 0.64$ V, $V_2 = 10.1$ V
2. $R_B = 0.9$ MΩ, $R_C = 2.6$ kΩ, $V_1 = 0.64$ V, $V_2 = 10.1$ V

$$I_{B1} = (20\text{ V} - 0.64\text{ V}) \div 1\text{ M}\Omega = 19.4\ \mu\text{A}$$
$$I_{C1} = (20\text{ V} - 10.1\text{ V}) \div 2.9\text{ k}\Omega = 3.41\text{ mA}$$
$$I_{B2} = (20\text{ V} - 0.64\text{ V}) \div 0.9\text{ M}\Omega = 21.5\ \mu\text{A}$$
$$I_{C2} = (20\text{ V} - 10.1\text{ V}) \div 2.6\text{ k}\Omega = 3.81\text{ mA}$$

$$\beta = \frac{\Delta I_C}{\Delta I_B} = \frac{3.81\text{ mA} - 3.41\text{ mA}}{21.5\ \mu\text{A} - 19.4\ \mu\text{A}} = 190$$

V. PROBLEMS

Sec. 9-1 Coupling Capacitor

9-1. For the circuit of Fig. 9-9, do the following:
   a. Find the critical frequency.
   b. Find the high-frequency border.
   c. What is the ac voltmeter reading when the generator's frequency is set to the critical frequency?
   d. What is the approximate ac voltmeter reading when the generator's frequency is set above the high-frequency border?

9-2. Repeat Prob. 9-1 for the circuit of Fig. 9-10.

9-3. For the circuit of Fig. 9-10, what must the value of the coupling capacitor be changed to for the critical frequency to become 0.1 Hz.

9-4. For the circuit of Fig. 9-11, do the following:
   a. Complete Table 9-1 for the given circuit conditions.

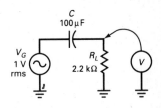

Figure 9-9

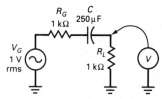

Figure 9-10

C μF	R_L kΩ	V_G V_{rms}	Critical Frequency
10	1	1	
100	1	1	
10	6.8	1	
10	1	10	

Table 9-1

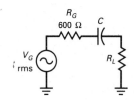

Figure 9-11

Select either *increased*, *decreased*, or *stayed the same* as the correct answer to the following questions.

   b. What happened to the critical frequency when only C was increased?
   c. What happened to the critical frequency when only R_L was increased?
   d. What happened to the critical frequency when only V_G was increased?

Sec. 9-2 Bypass Capacitors

9-5. For the circuit of Fig. 9-12, do the following:
   a. Find the critical frequency.
   b. Find the high-frequency border.
   c. What is the ac voltmeter reading when the generator's frequency is set to the critical frequency?
   d. What is the approximate ac voltmeter reading, when the generator's frequency is set above the high-frequency border?

9-6. Repeat Prob. 9-5 for the circuit of Fig. 9-13.

9-7. For the circuit of Fig. 9-13, what must the value of the coupling capacitor be changed to for the critical frequency to become 1 Hz.

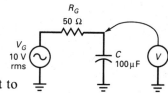

Figure 9-12

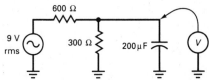

Figure 9-13

9-8. For the circuit of Fig. 9-14, do the following:
  a. Complete Table 9-2 for the given circuit conditions.

C μF	R_L kΩ	V_G V_{rms}	Critical Frequency
10	1	1	
100	1	1	
10	6.8	1	
10	1	10	

Table 9-2

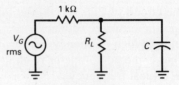

Figure 9-14

Select either *increased*, *decreased*, or *stayed the same* as the correct answer to the following questions.
  b. What happened to the critical frequency when only C was increased?
  c. What happened to the critical frequency when only R_L was increased?
  d. What happened to the critical frequency when only V_G was increased?

Sec. 9-3 Superposition in Amplifiers

9-9. For the circuit of Fig. 9-15, do the following:
  a. Draw the dc equivalent circuit. (Omit the portion of the circuit with the opened capacitors and the shorted ac source from the drawing.)
  b. Draw the ac equivalent circuit. (Combine all resistors that are in parallel.)
  c. Find I_E.
  d. Find V_{CE}.

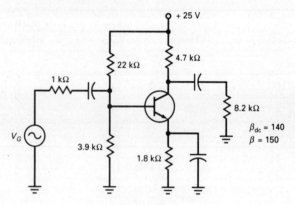

Figure 9-15

9-10. Repeat Prob. 9-9 for the circuit of Fig. 9-16.

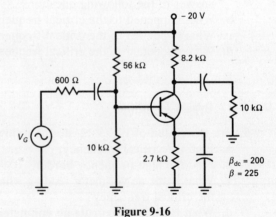

Figure 9-16

9-11. For the circuit of Fig. 9-17, do the following:
  a. Determine the readings of all the dc voltmeters.
  b. If voltmeter (V_2) were an ac voltmeter, what would be its reading?

Sec. 9-4 Small-Signal Operation

9-12. What is the maximum peak-to-peak value of the ac emitter current allowed in the circuit of Fig. 9-15, if the amplifier is to operate in the small-signal mode?

9-13. The peak-to-peak value of the ac emitter current for the circuit of Fig. 9-16 is 0.2 mA. Is this amplifier operating in the small-signal mode?

Sec. 9-5 AC Resistance of the Emitter Diode

9-14. A transistor amplifier has a dc current of 1 mA. What is the ac resistance of the emitter diode (r'_e)?

9-15. What is the ac resistance of the emitter diode (r'_e) in the circuit of Fig. 9-15?

9-16. From the dc voltmeter readings for the circuit of Fig. 9-18, determine the ac resistance of the emitter diode.

9-17. For the circuit of Fig. 9-19, do the following:
  a. Complete Table 9-3 for the given circuit conditions.

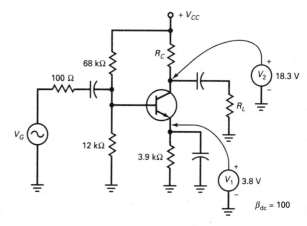

Figure 9-17

Figure 9-18

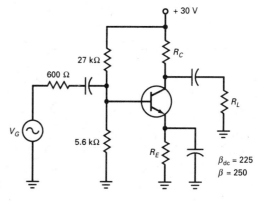

Figure 9-19

R_E kΩ	R_C kΩ	R_L kΩ	I_E mA	V_{CE} V	r'_e Ω
1.8	4.7	10			
3.3	4.7	10			
1.8	8.2	10			
1.8	4.7	22			

Table 9-3

Select either *increased*, *decreased*, or *stayed the same* as the correct answer to the following questions.
  b. What happened to r'_e when only R_E was increased?
  c. What happened to r'_e when only R_C was increased?
  d. What happened to r'_e when only R_L was increased?

Sec. 9-8 AC Model of a CE Amplifier

9-18. The circuit of Fig. 9-20 is the ac equivalent circuit of a common emitter amplifier. The ac resistance of the emitter diode is 25 Ω and the ac current gain is 200. Find the following:
  a. $z_{in(base)}$
  b. z_{in}

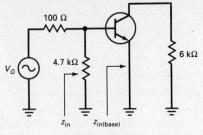

Figure 9-20

9-19. For the circuit of Fig. 9-15, do the following:
  a. Find $z_{in(base)}$ and z_{in} when $\beta = 150$.
  b. Find $z_{in(base)}$ and z_{in} when $\beta = 200$.
  c. What happened to the input impedance of the stage when only the current gain was increased?
  d. What would happen to the input impedance of the stage if only the emitter diode resistance was increased?

9-20. Find the input impedance of the stage in the circuit of Fig. 9-21.

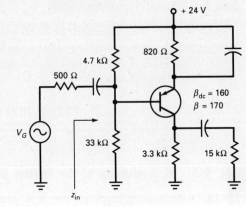

Figure 9-21

9-21. From the dc voltmeter readings for the circuit of Fig. 9-22, determine the input impedance of the stage.

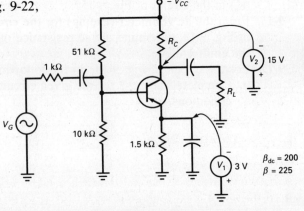

Figure 9-22

Sec. 9-9 AC Quantities on the Data Sheet

9-22. From the data of circuit measurements for the circuit of Fig. 9-23, determine the ac current gain of the transistor. Refer to the data sheet of the 2N3904 to see if this measured value is within specifications. Data:

1. $R_B = 1.93$ MΩ, $R_C = 10$ kΩ, $V_1 = 0.67$ V, $V_2 = 10$ V
2. $R_B = 1.76$ MΩ, $R_C = 8.85$ kΩ, $V_1 = 0.67$ V, $V_2 = 10$ V

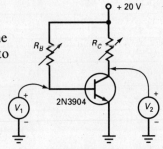

Figure 9-23

CHAPTER 10
VOLTAGE AMPLIFIERS

Study Chap. 10 in *Electronic Principles*.

I. TRUE / FALSE

Answer true (T) or false (F) to each of the following statements.   *Answer*

1. The ac collector current is approximately equal to the ac emitter current. **(10-1)**   1. ___
2. The ac collector voltage of a CE amplifier is in phase with the input voltage. **(10-1)**   2. ___
3. Due to the effect of the output coupling capacitor, the load voltage is a pure ac voltage. **(10-1)**   3. ___
4. The rms and peak-to-peak values of sinusoidal quantities are mathematically related. **(10-1)**   4. ___
5. When β increases, the input impedance of a CE amplifier will increase. **(10-1)**   5. ___
6. The voltage gain of an amplifier is defined as the ac output voltage divided by the generator voltage. **(10-2)**   6. ___
7. The ac input voltage of a CE amplifier is equal to the generator voltage, if the internal resistance of the generator is zero. **(10-2)**   7. ___
8. When β increases, the voltage gain of a CE amplifier will increase. **(10-3)**   8. ___
9. If the generator's voltage is increased, the voltage gain of a CE amplifier would increase. **(10-3)**   9. ___
10. In troubleshooting, answers within 10 to 20 percent are usually adequate. **(10-4)**   10. ___
11. The output of a swamped amplifier is less sensitive to the variations in the characteristics of the transistor. **(10-5)**   11. ___
12. The presence of a feedback resistor decreases the input impedance of an amplifier. **(10-5)**   12. ___
13. The presence of a feedback resistor decreases the voltage gain of an amplifier. **(10-5)**   13. ___
14. The total ac resistance in the emitter branch of a swamped amplifier is the parallel combination of r_e and r'_e. **(10-5)**   14. ___
15. If the feedback resistor of an amplifier swamps the nonlinear ac resistance of the emitter diode, it indirectly reduces the distortion that occurs for large-signal operation. **(10-5)**   15. ___
16. The emitter of a swamped amplifier is at ac ground. **(10-5)**   16. ___
17. The ac collector resistance of the first stage in a two-stage amplifier includes the input impedance of the second stage. **(10-6)**   17. ___

18. If three swamped amplifiers were cascaded and each had a voltage gain of 10, the total voltage gain of the three-stage amplifier would be 30. **(10-6)** 18. ___
19. If the collector resistor of a CE amplifier is shorted, the ac output voltage would equal zero. **(10-7)** 19. ___
20. If the emitter resistor of a CE amplifier is opened, the ac output voltage would equal zero. **(10-7)** 20. ___

II. COMPLETION
Complete each of the following statements.   *Answer*

1. In a CE amplifier the generator voltage is coupled through the input capacitor into the ___ of the transistor. **(10-1)** 1. ___
2. When the emitter of a CE amplifier is at ac ground, all the ac base voltage appears across the ___ diode. **(10-1)** 2. ___
3. One shortcoming of the CE amplifier is that the input impedance is ___ sensitive. **(10-1)** 3. ___
4. The Δ quantities are always ___ quantities. **(10-1)** 4. ___
5. Lowercase letters are used to represent ___ voltages and currents. **(10-1)** 5. ___
6. The voltage gain shows how much ___ the circuit provides. **(10-2)** 6. ___
7. The voltage gain times the ac input voltage equals the ac ___ voltage. **(10-2)** 7. ___
8. The predicted voltage gain of a CE amplifier is equal to r_c divided by ___ for a small-signal operation. **(10-3)** 8. ___
9. If the load resistor decreases, the voltage gain will ___. **(10-3)** 9. ___
10. If a CE amplifier is working correctly, the measured voltage gain will approximately equal the ___ voltage gain. **(10-3)** 10. ___
11. The portion of the emitter resistance that is left unbypassed is called a ___ resistor. **(10-5)** 11. ___
12. The purpose of using a partially bypass emitter resistance is to ___ the voltage gain. **(10-5)** 12. ___
13. The condition where r_e is much larger than r'_e and, as a result, changes in r'_e are greatly reduced in effect can be stated as r_e ___ r'_e. **(10-5)** 13. ___
14. The total ac resistance in the collector branch of a swamped amplifier is the parallel combination of R_C and ___. **(10-5)** 14. ___
15. When troubleshooting a swamped amplifier, for a quick gain estimate, you can ignore the effects of ___. **(10-5)** 15. ___
16. The coupling of the output of one stage to the input of another stage is called ___ the stages. **(10-6)** 16. ___
17. The total voltage gain of a two-stage amplifier is the ___ of the gains of each stage. **(10-6)** 17. ___
18. If the load resistance of a CE amplifier is opened, the value of the ac output voltage would ___. **(10-7)** 18. ___
19. If the bypass capacitor of a CE amplifier is opened, the value of the ac output voltage would ___. **(10-7)** 19. ___
20. If the output capacitor of a CE amplifier is shorted, the value of the dc collector voltage would ___. **(10-7)** 20. ___

III. CIRCUIT HIGHLIGHTS

A. AC Equivalent Circuit of a CE Amplifier with a T Model Transistor

B. AC Equivalent Circuit of a CE Amplifier with a Feedback Resistor and a T Model Transistor

(Swamped Amplifier, When $r_e > r'_e$)

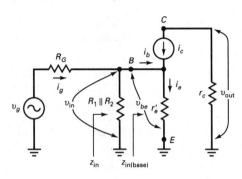

Figure 10-1

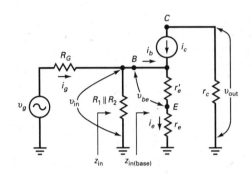

Figure 10-2

$z_{in(base)} = \beta r'_e$

$z_{in} = R_1 // R_2 // \beta r'_e$

$v_{in} = \dfrac{z_{in}}{z_{in} + R_G} v_g$

$v_{in} = v_{be}$

$v_{be} = i_e r'_e$

$A = \dfrac{r_c}{r'_e}$

$v_{out} = A v_{in}$

$i_c \approx i_e$

$i_b = \dfrac{i_c}{\beta}$

$i_c = \dfrac{v_{out}}{r_c}$

$i_g = \dfrac{v_g}{R_G + z_{in}}$

$z_{in(base)} = \beta(r'_e + r_e)$

$z_{in} = R_1 // R_2 // \beta(r'_e + r_e)$

$v_{in} = \dfrac{z_{in}}{z_{in} + R_G} v_g$

$v_{in} = v_{be} + i_e r_e$

$v_{be} = i_e r'_e$

$A = \dfrac{r_c}{r_e + r'_e}$

$v_{out} = A v_{in}$

$i_c \approx i_e$

$i_b = \dfrac{i_c}{\beta}$

$i_c = \dfrac{v_{out}}{r_c}$

$i_g = \dfrac{v_g}{R_G + z_{in}}$

IV. ILLUSTRATIVE AND PRACTICE PROBLEMS

Illustrative Problem 1. For the CE amplifier, find the following:
a. v_{in}
b. i_e
c. v_{out}
d. A
e. A (predicted)

Note: All ac currents and voltages will be in peak-to-peak values, unless otherwise indicated.

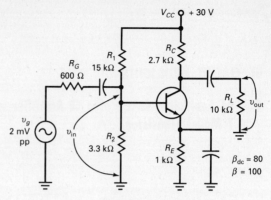

Figure 10-3

Steps	Comments
1. Draw the ac equivalent circuit. **Figure 10-4**	a. This ac equivalent circuit includes the model T. b. An earlier analysis of this circuit was done in Chapter 9 and yielded the following results: $I_E = 4.71$ mA $\quad r'_e = 5.31\ \Omega$ $r_c = 2.13$ kΩ $\quad z_{in} = 444\ \Omega$ $R_1 // R_2 = 2.7$ kΩ
2. Find v_{in}. $v_{in} = \dfrac{444\ \Omega}{444\ \Omega + 600\ \Omega}\, 2\text{ mV} = 0.851$ mV	Voltage divider rule. $v_{in} = \dfrac{z_{in}}{z_{in} + R_G}\, v_g$
3. Find i_e. $i_e = \dfrac{0.851\text{ mV}}{5.31\ \Omega} = 0.160$ mA	Emitter diode dynamics. $i_e = \dfrac{v_{be}}{r'_e} = \dfrac{v_{in}}{r'_e}$ $\quad$ substitution $v_{be} = v_{in}$
4. Find v_{out}. $v_{out} = (0.160\text{ mA})(2.13\text{ k}\Omega) = 341$ mV	Ohm's law. $v_{out} = i_c r_c = i_e r_c$ $\quad$ substitution $i_c \approx i_e$
5. Find A. $A = \dfrac{341\text{ mV}}{0.851\text{ mV}} = 401$	Definition of the voltage gain. $A = \dfrac{v_{out}}{v_{in}}$
6. Find A (predicted). $A = \dfrac{2.13\text{ k}\Omega}{5.31\ \Omega} = 401$	Voltage gain derived from circuit conditions. $A = \dfrac{r_c}{r'_e}$

Name _____ Date _____

Practice Problem 1. For the CE amplifier, find the following:
a. v_{in} b. i_e
c. v_{out} d. A
e. A (predicted)
(This circuit was a practice problem in Chapter 9 and should have yielded the following results):

$I_E = 2.32$ mA $r'_e = 10.8\ \Omega$ $z_{in} = 1.21$ kΩ

$r_c = 2.94$ kΩ $R_1//R_2 = 4.79$ kΩ

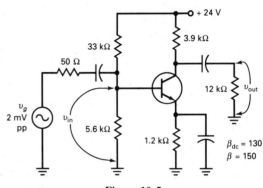

Figure 10-5

Answers:
a. $v_{in} = 1.92$ mV   b. $i_e = 0.178$ mA   c. $v_{out} = 523$ mV   d. $A = 272$ e. $A = 272$ (predicted)

Illustrative Problem 2. For the swamped amplifier, find the following. (The circuit in Illustrative Problem 1 has been modified to make it a swamped amplifier.)
a. z_{in} b. A
c. v_{in} d. v_{out}

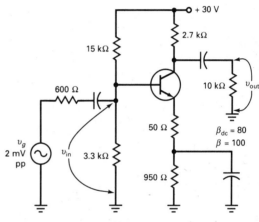

Figure 10-6

Steps	Comments
1. Draw the ac equivalent circuit. Figure 10-7	a. The emitter resistance of 1000 Ω has been partially bypassed. The values for all the dc voltages and currents will not change. See Illustrative Problem 1. $I_E = 4.71$ mA $r'_e = 5.31\ \Omega$ b. The following quantities will change: $z_{in(base)}$, z_{in}, A, and all the ac voltages and currents.
2. Find $z_{in(base)}$. $z_{in(base)} = 100(50\ \Omega + 5.31\ \Omega) = 5.53$ kΩ	a. Input impedance of the base. b. The portion of R_E that is not bypassed is r_e. $z_{in(base)} = \beta(r_e + r'_e)$

Copyright © 1993 by the Glencoe Division of Macmillan/McGraw-Hill School Publishing Company. All rights reserved.

Steps	Comments
3. Find z_{in}. $z_{in} = \dfrac{(2.7\ k\Omega)(5.53\ k\Omega)}{(2.7\ k\Omega + 5.53\ k\Omega)} = 1.81\ k\Omega$	Input impedance of the stage. $z_{in} = R_1 // R_2 // \beta(r_e + r'_e)$
4. Find A (predicted). $A = \dfrac{2.13\ k\Omega}{(50\ \Omega + 5.31\ \Omega)} = 38.5$	Voltage gain derived from circuit conditions. $A = \dfrac{r_c}{r_e + r'_e}$
5. Find v_{in}. $v_{in} = \dfrac{1.81\ k\Omega}{1.81\ k\Omega + 600\ \Omega}\ 2\ mV = 1.5\ mV$	Voltage divider rule. $v_{in} = \dfrac{z_{in}}{z_{in} + R_G}\ v_g$
6. Find v_{out}. $v_{out} = (38.5)(1.5\ mV) = 57.8\ mV$	From the voltage gain formula. $v_{out} = Av_{in}$

Practice Problem 2. For the swamped amplifier, find the following. (The circuit in Practice Problem 1 has been modified to make it a swamped amplifier without changing any of its dc values.)

a. z_{in}
b. A
c. v_{in}
d. v_{out}

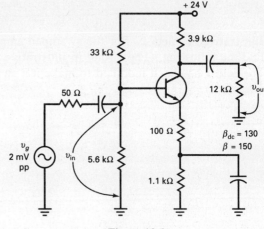

Figure 10-8

Answers:
a. $z_{in} = 3.72\ k\Omega$ b. $A = 26.5$ c. $v_{in} = 1.97\ mV$ d. $v_{out} = 52.2\ mV$

V. PRACTICAL TECHNIQUES

Objective: Measure the input impedance of an amplifier.

Steps	Comments
1. Construct the circuit that is shown in Fig. 10-9. **Figure 10-9**	a. The sensing resistor (R_{sen}) is used so that the generator current can be measured with an oscilloscope. b. In order to make voltage measurements that are discernible, choose a sensing resistor whose value is not much greater or smaller than the expected value of z_{in}. c. For a CE amplifier, choose a value of R_{sen} that is between 1 and 10 kΩ. d. Set the frequency of the voltage generator to about 10 kHz.
2. Measure v_a.	Adjust the voltage generator so that v_a measures 10-mV peak-to-peak on the oscilloscope.
3. Measure v_b.	a. Move the oscilloscope lead from node A to node B. b. Measure the voltage at node B. c. Record the reading.
4. Determine i_g. $$i_g = \frac{v_a - v_b}{R_{sen}}$$	a. The generator current is the input current to the amplifier and can be determined from applying Ohm's law. b. Calculate i_g from measured values.
5. Determine z_{in}. $$z_{in} = \frac{v_b}{i_g}$$	a. The input impedance is equal to the input voltage v_b divided by the input current i_g. b. Calculate z_{in} from measured values.

Example 1. From the data of circuit measurements for the circuit of Fig. 10-10, determine the input impedance of the CE amplifier.

Data:

$$v_a = 10 \text{ mV} \quad v_b = 2 \text{ mV}$$

$$i_g = \frac{v_a - v_b}{R_{sen}} = \frac{10 \text{ mV} - 2 \text{ mV}}{10 \text{ k}\Omega} = 0.8 \text{ }\mu\text{A}$$

$$z_{in} = \frac{v_b}{i_g} = \frac{2 \text{ mV}}{0.8 \text{ }\mu\text{A}} = 2.5 \text{ k}\Omega$$

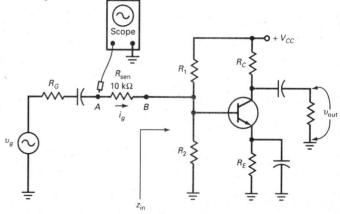

Figure 10-10

VI. PROBLEMS

(*Note:* Assume $\beta_{dc} = 175$ and $\beta = 200$ in all problems unless otherwise indicated.)
All ac currents and voltages will be in peak-to-peak values unless otherwise indicated.

Sec. 10-1 Highlights of a CE Amplifier

10-1. The circuit of Fig. 10-11 is an ac equivalent circuit of a CE amplifier with a T model transistor. The ac resistance of the emitter diode is 20 Ω.
  a. Find $z_{in(base)}$.
  b. Find z_{in}.

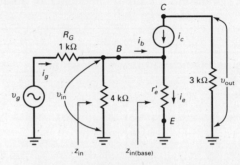

Figure 10-11

10-2. The circuit of Fig. 10-12 is an ac equivalent circuit of a CE amplifier with a Π model transistor. The ac resistance of the emitter diode is 12 Ω.
  a. Find $z_{in(base)}$.
  b. Find z_{in}.

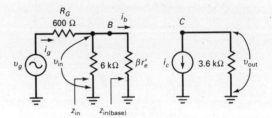

Figure 10-12

10-3. The circuit of Fig. 10-13 is an ac equivalent circuit of a CE amplifier. The ac resistance of the emitter diode is 10 Ω.
  a. Find $z_{in(base)}$.
  b. Find z_{in}.

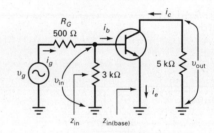

Figure 10-13

10-4. Find the input impedance of the CE amplifier in Fig. 10-14.

* **10-5.** From the data of circuit measurements for the circuit of Fig. 10-15, determine the input impedance of the CE amplifier. Data:

$$v_a = 10 \text{ mV} \qquad v_b = 4 \text{ mV}$$

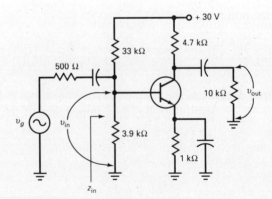

Figure 10-14

* See "Practical Techniques."

Sec. 10-2 Voltage Gain

10-6. From the data of circuit measurements for the circuit of Fig. 10-15, determine the voltage gain of the amplifier. Data:

$$v_b = 4 \text{ mV} \qquad v_{out} = 0.6 \text{ V}$$

10-7. With respect to Prob. 10-1, when $v_g = 1$ mV, find the following:
- a. v_{in}
- b. i_g
- c. v_{be}
- d. i_e
- e. v_{out}
- f. A

10-8. With respect to Prob. 10-2, when $v_g = 2$ mV, find the following:
- a. v_{in}
- b. i_b
- c. i_c
- d. v_{out}
- e. A

10-9. For Fig. 10-13, when $v_g = 2$ mV and $r'_e = 15\ \Omega$, find the following:
- a. v_{in}
- b. v_{out}
- c. A

10-10. With respect to Prob. 10-4, when $V_g = 3$ mV, find the following:
- a. v_{in}
- b. i_g
- c. i_c
- d. v_{out}
- e. A

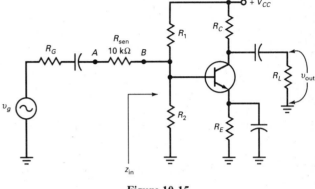

Figure 10-15

Sec. 10-3 Predicting Voltage Gain

10-11. The circuit of Fig. 10-16 is an equivalent circuit of a CE amplifier with a T model transistor. What is the voltage gain?

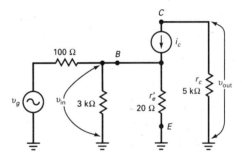

Figure 10-16

10-12. The dc emitter current is 1 mA in the circuit of Fig. 10-17. What is the voltage gain?

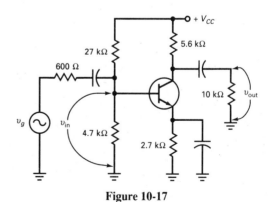

Figure 10-17

10-13. For the ac equivalent circuit of the CE amplifier in Fig. 10-18, do the following:
   a. Complete Table 10-1 for the given circuit conditions.

Table 10-1

r'_e, Ω	R_L, kΩ	β	v_{in}, mV	A
15	3.3	200		
30	3.3	200		
15	10	200		
15	3.3	400		

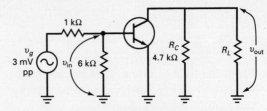

Figure 10-18

Select either *increased*, *decreased*, or *stayed the same* as the correct answer to the following questions.

b. What happened to the input voltage when only r'_e was increased?
c. What happened to the input voltage when only R_L was increased?
d. What happened to the input voltage when only β was increased?
e. What happened to the voltage gain when only r'_e was increased?
f. What happened to the voltage gain when only R_L was increased?
g. What happened to the voltage gain when only β was increased?

10-14. For the circuit of Fig. 10-19, what is the output voltage?

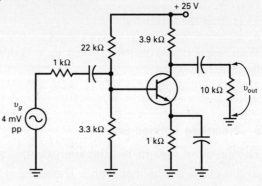

Figure 10-19

10-15. For the circuit of Fig. 10-20, what is the output voltage?

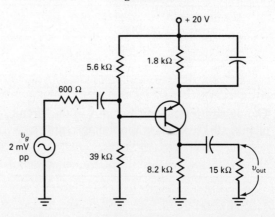

Figure 10-20

Sec. 10-5 Swamped Amplifier

10-16. The circuit of Fig. 10-21 is an ac equivalent circuit of a swamped amplifier with a T model transistor. Find the following:
  a. $Z_{in(base)}$ **b.** Z_{in}
  c. v_{in} **d.** A
  e. v_{out}

10-17. For the ac equivalent circuit of the swamped amplifier in Fig. 10-22 whose emitter diode ac resistance is 10 Ω, do the following:
  a. Complete Table 10-2 for the given circuit conditions.

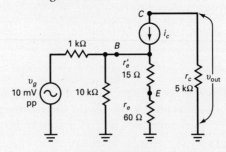

Figure 10-21

r_e, Ω	R_L, kΩ	β	v_{in}, mV	A
50	3.3	100		
100	3.3	100		
50	10	100		
50	3.3	200		

Table 10-2

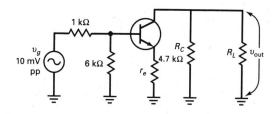

Figure 10-22

Select either *increased*, *decreased*, or *stayed the same* as the correct answer to the following questions.

b. What happened to the input voltage when only r_e was increased?
c. What happened to the input voltage when only R_L was increased?
d. What happened to the input voltage when only β was increased?
e. What happened to the voltage gain when only r_e was increased?
f. What happened to the voltage gain when only R_L was increased?
g. What happened to the voltage gain when only β was increased?

10-18. For the circuit of Fig. 10-23, find the following:
  a. z_{in} **b.** v_{in}
  c. A   **d.** v_{out}

10-19. For the circuit of Fig. 10-24, what is the output voltage?

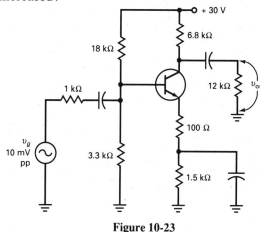

Figure 10-23

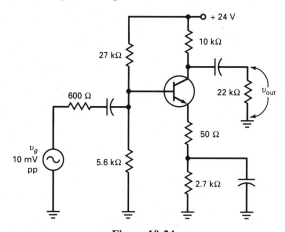

Figure 10-24

Sec. 10-16 Cascaded Stages

10-20. For the circuit of Fig. 10-25, find the following:
  a. v_{in} (first stage)
  b. v_{in} (second stage)
  c. A (first stage)
  d. A (second stage)
  e. A (total gain)
  f. v_{out}

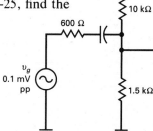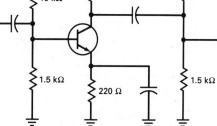

Figure 10-25

10-21. If β = 400 for the transistors of Fig. 10-25, what is the ac output voltage of the second stage?

10-22. For the circuit of Fig. 10-26, find the following:
- **a.** v_{in} (first stage)
- **b.** v_{in} (second stage)
- **c.** A (first stage)
- **d.** A (second stage)
- **e.** A (total gain)
- **f.** v_{out}

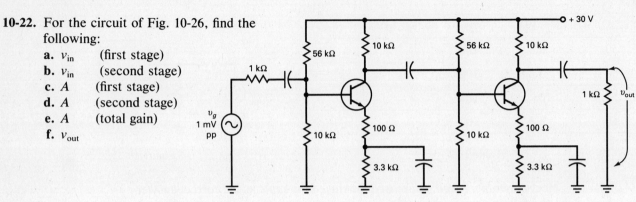

Figure 10-26

10-23. If β = 400 for the transistors of Fig. 10-26, what is the ac output voltage of the second stage?

Sec. 10-7 Troubleshooting

10-24. For the circuit of Fig. 10-27, fill in the readings of the oscilloscope in Table 10-3 for the given circuit conditions.

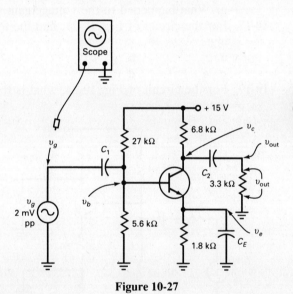

Figure 10-27

Trouble	Oscilloscope Readings				
	v_g, mV	v_b, mV	v_c, mV	v_e, mV	v_{out}, mV
Circuit OK					
C_E opened					
C_1 opened					
C_2 opened					
Voltage generator off					
DC power supply off					

Table 10-3

CHAPTER 11
POWER AMPLIFIERS

Study Chap. 11 in *Electronic Principles*.

I. TRUE / FALSE

Answer true (T) or false (F) to each of the following statements.

Answer

1. Every amplifier has a dc load line and an ac load line. (11-1)
2. The operating point of an amplifier moves along the dc load line. (11-1)
3. When the ac collector resistance is different from the dc collector resistance, the ac load line will be different from the dc load line. (11-1)
4. In analyzing biasing circuits, ac load lines are used. (11-1)
5. Clipping is undesirable because it results in excessive distortion of signals. (11-2)
6. Clipping occurs when the input signal is too large. (11-2)
7. The three key quantities that are needed to determine the maximum peak-to-peak output voltage of a fully bypassed CE amplifier are I_{CQ}, V_{CEQ}, and r_c. (11-2)
8. If the Q point is at the center of the dc load line, clipping is more likely to occur simultaneously on both the positive and negative peaks of the output voltage. (11-2)
9. If the Q point is below the center of the dc load line, you will get $I_{CQ}r_c$ clipping first. (11-2)
10. If the load resistance is varied, the Q point on the ac load line will shift. (11-2)
11. Clipping occurs when the ac collector voltage tries to exceed the upper and lower limits that are imposed on the transistor characteristics by the ac load line. (11-2)
12. Class A operation means that the transistor operates in the active region at all times. (11-3)
13. The power dissipation of the transistor is increased when an ac signal is applied to the amplifier. (11-3)
14. Maximum load power occurs in an amplifier when the amplifier is producing the maximum peak-to-peak unclipped output. (11-3)
15. The total dc power supplied to an amplifier is the product of the dc supply voltage and the collector current. (11-3)
16. The efficiency of an amplifier is the ac load power divided by the total dc power that is supplied by the power supply and then multiplied by 100 percent. (11-3)
17. The power dissipation of a transistor must never exceed the power rating of the transistor. (11-3)
18. If the load voltage is doubled, the load power will also double. (11-3)

19. When the ambient temperature increases, the power rating of the transistor will increase. (11-4)

19. _____

20. Data sheets specify the maximum junction temperatures as $T_{j(max)}$. (11-4)

20. _____

II. COMPLETION

Complete each of the following statements.

Answer

1. The Q point is where the ac and dc load lines _____. (11-1)
2. The quiescent collector-emitter voltage is the same as the _____ collector-emitter voltage. (11-1)
3. Large signal operations are analyzed by using the _____ load line. (11-1)
4. When the ac collector resistance is less than the dc collector resistance, the slope of the ac load line will be _____ than the slope of the dc load line. (11-1)
5. The symbol that is used for the maximum peak-to-peak unclipped ac output voltage is _____. (11-2)
6. In a fully bypassed CE amplifier, the ac collector-emitter voltage is equal to the _____ voltage. (11-2)
7. When the Q point is centered on the ac load line, $I_{CQ}r_c$ equals _____. (11-2)
8. The maximum unclipped peak-to-peak output voltage is equal to the smaller of $2I_{CQ}r_c$ or _____. (11-2)
9. The maximum peak-to-peak unclipped ac output voltage is optimized when the Q point is centered on the _____ load line. (11-2)
10. In class A operation, the number of degrees that the collector current flows of the ac cycle is _____ degrees. (11-3)
11. The power dissipated by the load is also called the _____ power. (11-3)
12. When peak-to-peak values are used in the calculation of power, the number _____ will appear in the divisor. (11-3)
13. When there is no input signal to an amplifier, the power dissipation of the transistor equals the product of V_{CEQ} and _____. (11-3)
14. The ac output power divided by the ac input power is called the power _____ of the amplifier. (11-3)
15. The symbol used for the total dc power supplied to the amplifier is _____. (11-3)
16. The theoretical limit for the efficiency of a class A amplifier is _____ percent. (11-3)
17. The sum of the voltage divider current and the collector current is called the current _____ of the stage. (11-3)
18. The symbol used for the efficiency of an amplifier is _____. (11-3)
19. The temperature of the surrounding air is known as the _____ temperature. (11-4)
20. The device that is fastened to a transistor to get rid of heat faster is called a _____ sink. (11-4)

1. _____
2. _____
3. _____
4. _____
5. _____
6. _____
7. _____
8. _____
9. _____
10. _____
11. _____
12. _____
13. _____
14. _____
15. _____
16. _____
17. _____
18. _____
19. _____
20. _____

III. ILLUSTRATIVE AND PRACTICE PROBLEMS

Illustrative Problem 1. Find the maximum generator voltage that produces an unclipped output signal.
Note: All the ac currents and voltages in this chapter will be in peak-to-peak values, unless otherwise indicated.

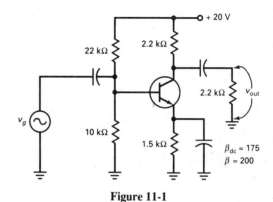

Figure 11-1

Steps	Comments
1. Draw the ac load line. (Optional) Figure 11-2	a. The ac and dc analysis that was done on this circuit yielded the following results: $I_E = 3.7$ mA $V_{CE} = 6.31$ V $r'_e = 6.76\ \Omega$ $z_{in} = 1.13$ kΩ $r_c = 1.1$ kΩ $A = 163$ b. The ac load line is drawn to help illustrate how the MPP is found.
2. Find V_{CEQ}. $V_{CEQ} = 6.31$ V	$V_{CE} = V_{CEQ}$
3. Find $I_{CQ}r_c$. $I_{CQ}r_c = (3.7$ mA$)(1.1$ k$\Omega) = 4.07$ V	$I_E \approx I_{CQ}$
4. Find MPP. MPP $= 2(3.7$ mA$)(1.1$ k$\Omega) = 8.14$ V	a. If $V_{CEQ} < I_{CQ}r_c$, then MPP $= 2V_{CEQ}$. b. If $V_{CEQ} > I_{CQ}r_c$, then MPP $= 2I_{CQ}r_c$. Since $V_{CEQ} > I_{CQ}r_c$, therefore MPP $= 2I_{CQ}r_c$.
5. Find v_{in} (maximum). $v_{in} = \dfrac{8.14\text{ V}}{163} = 49.9$ mV	The maximum v_{in} for an unclipped output can be determined by using the value of MPP in the gain formula. $v_{in} = \dfrac{v_{out}}{A} = \dfrac{\text{MPP}}{A}$
6. Find v_g (maximum). $v_g = 49.9$ mV	Since $R_G = 0\ \Omega$, $v_{in} = v_g$.

Practice Problem 1. Find the maximum generator voltage that produces an unclipped output signal. The ac and dc analysis that was done on this circuit yielded the following results:

$I_E = 7.17 \text{ mA} \quad V_{CE} = 5.21 \text{ V}$
$r'_e = 3.49 \text{ }\Omega \quad r_C = 0.818 \text{ k}\Omega$
$A = 234$

Answer: $v_{in} = 44.4$ mV

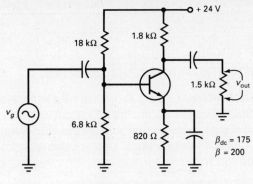

Figure 11-3

Illustrative Problem 2. For the circuit of Illustrative Problem 1, do the following:
a. Find the maximum efficiency of the amplifier.
b. Find the efficiency of the amplifier when v_{out} is equal to 0.5 MPP.

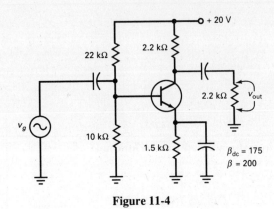

Figure 11-4

Steps	Comments
1. Find I_1. $$I_1 = \frac{20 \text{ V}}{22 \text{ k}\Omega + 10 \text{ k}\Omega} = 0.625 \text{ mA}$$	a. Neglect the base current because it is very small. b. Apply Ohm's law. $$I_1 = \frac{V_{CC}}{R_1 + R_2}$$
2. Find I_S. $$I_S = 0.625 \text{ mA} + 3.7 \text{ mA} = 4.33 \text{ mA}$$	a. The total drain current of the power supply is equal to $I_1 + I_C$. b. Illustrative Problem 1 yielded the following results: $$I_C = 3.7 \text{ mA} \quad \text{MPP} = 8.14 \text{ V}$$
3. Find P_S. $$P_S = (20 \text{ V})(4.33 \text{ mA}) = 86.6 \text{ mW}$$	The total dc power supplied to the amplifier by the power supply is equal to V_{CC} times I_S.
4. Find P_L (maximum). $$P_L = \frac{(8.14 \text{ V})^2}{(8)(2.2 \text{ k}\Omega)} = 3.76 \text{ mW}$$	Maximum power delivered to the load occurs when v_{out} is equal to MPP. $$P_L = \frac{v_{out}^2}{8R_L} = \frac{\text{MPP}^2}{8R_L}$$
5. Find η (maximum). $$\eta = \frac{3.76 \text{ mW}}{86.6 \text{ mW}} 100\% = 4.34\%$$	Maximum efficiency occurs when the maximum power is delivered to the load. $$\eta = \frac{P_L}{P_S} 100\%$$

Steps	Comments
6. Find P_L. ($v_{out} = 4.07$ V) $$P_L = \frac{(4.07 \text{ V})^2}{(8)(2.2 \text{ k}\Omega)} = 0.941 \text{ mW}$$	Load power when v_{out} equals 0.5 MPP. $$P_L = \frac{v_{out}^2}{8R_L}$$
7. Find η. ($v_{out} = 4.07$ V) $$\eta = \frac{0.941 \text{ mW}}{86.6 \text{ mW}} (100\%) = 1.09\%$$	Efficiency when v_{out} equals 0.5 MPP. $$\eta = \frac{P_L}{P_S} 100\%$$

Practice Problem 2. For the circuit of Practice Problem 1, do the following. (Practice Problem 1 should have yielded the following results: $I_{CQ} = 7.17$ mA, MPP $= 10.4$ V.)
a. Find the maximum efficiency of the amplifier.
b. Find the efficiency of the amplifier when v_{out} is equal to 0.5 MPP.

Answers: $\eta = 4.62\%$ (maximum), $\eta = 1.15\%$

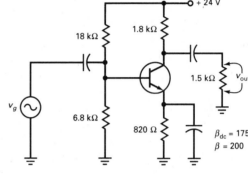

Figure 11-5

IV. PROBLEMS

Note: Assume $\beta_{dc} = 175$ and $\beta = 200$ for all problems in this chapter, unless otherwise indicated.

Sec. 11-2 Limits on Signal Swing

11-1. For the circuit and its ac load line of Fig. 11-6, find the following:
  a. MPP
  b. R_C
  c. r_c
  d. R_L

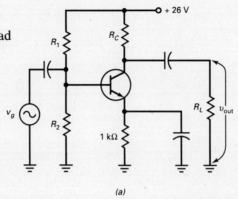
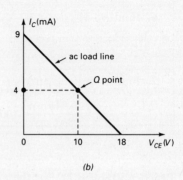

Figure 11-6

11-2. For the circuit of Fig. 11-7, do the following:
  a. Find V_{CEQ}.
  b. Find $I_{CQ}r_c$.
  c. Is the Q point above or below the ac load line's midpoint?
  d. Find MPP.
  e. If the generator voltage is 60 mV, is the output signal clipped?

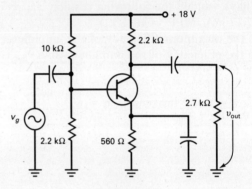

Figure 11-7

11-3. From the dc voltmeter reading for the circuit of Fig. 11-8, do the following:
  a. Find MPP.
  b. Find the maximum generator voltage that produces an unclipped output signal.

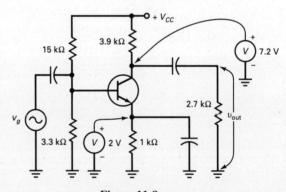

Figure 11-8

11-4. For the circuit of Fig. 11-9, find the maximum generator voltage that produces an unclipped output signal when $R_G = 0\ \Omega$ and $R_G = 1\ k\Omega$.

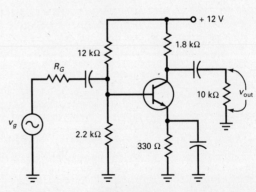

Figure 11-9

11-5. For the circuit of Fig. 11-10, find the maximum generator voltage that produces an unclipped output signal.

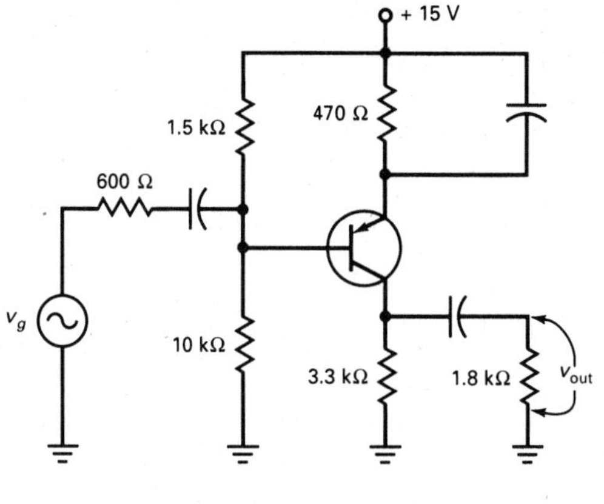

Sec. 11-3 Class A Operations

11-6. For the circuit of Fig. 11-7, find the following:
  a. Current drain.
  b. Total dc power supplied.
  c. Maximum power delivered to the load.
  d. Power dissipation of the transistor without any signal.
  e. Maximum efficiency.

11-7. For the circuit of Fig. 11-9, when $R_G = 0\ \Omega$, find the following:
  a. I_S
  b. P_S
  c. P_L (maximum)
  d. P_L (when $v_g = 40$ mV)
  e. η (maximum)
  f. η (when $v_g = 40$ mV)

11-8. For the circuit of Fig. 11-10, find the following:
  a. I_S
  b. P_S
  c. P_L (maximum)
  d. P_L (when $v_g = 80$ mV)
  e. η (maximum)
  f. η (when $v_g = 80$ mV)

Figure 11-10

Sec. 11-4 Transistor Power Rating

11-9. For the transistor power derating curve of Fig. 11-11, find the following:
  a. Maximum power rating at room temperature.
  b. Maximum power rating when the ambient temperature is 100°C.
  c. Derating factor (slope of curve).

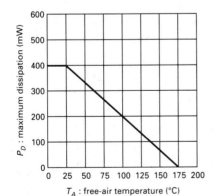

Figure 11-11

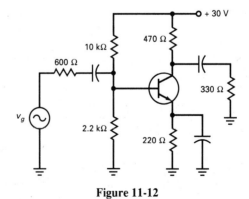

Figure 11-12

11-10. If the ambient temperature is 75°C for the circuit of Fig. 11-12, is the transistor in danger of overheating when no ac signal is applied? Use the transistor power derating curve of Fig. 11-11.

CHAPTER 12
EMITTER FOLLOWERS

Study Chap. 12 in *Electronic Principles*.

I. TRUE / FALSE

Answer true (T) or false (F) to each of the following statements.

Answer

1. There is no ac voltage on the emitter of an emitter follower. (12-1)   1. ___
2. Common-collector amplifiers are called emitter followers because the output voltage is in phase with the input voltage and has approximately the same peak value. (12-1)   2. ___
3. A high external ac emitter resistance of an emitter follower causes the voltage gain to be very stable, minimizes distortion, and produces a high-input impedance. (12-1)   3. ___
4. The input impedance at the base of an emitter follower will not be affected by any changes in the ac current gain. (12-2)   4. ___
5. The input impedance at the base of an emitter follower will not be affected by any changes in the load resistance. (12-2)   5. ___
6. The voltage gain of an emitter follower can be greater than one. (12-3)   6. ___
7. The input voltage of an emitter follower is slightly larger than the generator voltage. (12-3)   7. ___
8. The heavy swamping of an emitter follower virtually eliminates the effects of r'_e on the voltage gain. (12-3)   8. ___
9. The output voltage of an emitter follower is approximately equal to the input voltage. (12-3)   9. ___
10. A square wave output from an emitter follower implies that there is clipping only at saturation. (12-4)   10. ___
11. The MPP is equal to twice the smaller limit the output voltage can swing from the Q point on the ac load line. (12-4)   11. ___
12. The optimum Q point exists when it is centered on the dc load line. (12-4)   12. ___
13. The dc load line determines the limits of the output voltage swing. (12-4)   13. ___
14. An emitter follower can never be directly coupled to a CE amplifier. (12-5)   14. ___
15. When an emitter follower is coupled to the output of a CE amplifier, the input impedance of the emitter follower becomes the load resistance of the CE amplifier. (12-5)   15. ___
16. The main advantage of a Darlington connection is the high-input impedance looking into the base of the first transistor. (12-6)   16. ___

17. The TP101 is a Darlington transistor with a maximum β of 20,000. (12-6) 17. ___
18. The advantage of class B amplifiers is lower current drain and higher efficiency. (12-7) 18. ___
19. In class B operation, the Q point is located approximately at saturation on both the dc and ac load lines. (12-7) 19. ___
20. In class B operation, the design arrangement where one transistor conducts during one half-cycle and the other transistor conducts during the other half-cycle is called push-pull. (12-7) 20. ___

II. COMPLETION

Complete each of the following.

Answer

1. An amplifier whose collector is at ac ground is known as a ____-____ amplifier. (12-1) 1. _____
2. The capacitor in an emitter follower that couples the ac emitter voltage to the load resistor is called the ____ capacitor. (12-1) 2. _____
3. Because of the high external emitter resistance in an emitter follower, ____ feedback is very pronounced. (12-1) 3. _____
4. The external ac emitter resistance (r_e) of an emitter follower is equal to the parallel combination of R_E and ____. (12-2) 4. _____
5. The total ac emitter resistance of an emitter follower is equal to the series combination of r_e and ____. (12-2) 5. _____
6. When r_e is much greater than r'_e in an emitter follower, the voltage gain is approximately equal to ____. (12-3) 6. _____
7. The main use of the emitter follower is in isolating a ____ impedance load from a CE amplifier. (12-3) 7. _____
8. When a CE amplifier is said to be overloaded, it means that the load it is driving is too ____. (12-3) 8. _____
9. The output voltage is taken across the ____ resistor. (12-3) 9. _____
10. The maximum voltage swing from the left side of the Q point on the ac load line of an emitter follower is equal to ____. (12-4) 10. _____
11. The maximum voltage swing from the right side of the Q point on the ac load line of an emitter follower is equal to ____. (12-4) 11. _____
12. The limits of the output voltage swing of an emitter follower are similar to those of a CE amplifier, except that we need to use ____ instead of r_c for its determination. (12-4) 12. _____
13. If the input voltage to an emitter follower is too large, the output voltage will be ____. (12-4) 13. _____
14. When the signal between amplifier stages is coupled with only a wire, the stages are said to be ____ coupled. (12-5) 14. _____
15. When the signal between amplifier stages is coupled with a capacitor, the stages are said to be ____ coupled. (12-5) 15. _____
16. Two cascaded emitter followers are known as a ____ connection. (12-6) 16. _____
17. The dc base-emitter voltage of a Darlington transistor is about ____ V. (12-6) 17. _____
18. Class B operation of a transistor means that the collector current flows for only ____ degrees of the ac cycle. (12-7) 18. _____
19. The temperature problem that occurs when increases in temperature cause increases in collector current and vice versa is called thermal ____. (12-7) 19. _____
20. The external diodes that are used in a voltage divider class B amplifier that compensate for changes in temperature are called ____ diodes. (12-7) 20. _____

III. ILLUSTRATIVE AND PRACTICE PROBLEMS

Illustrative Problem 1. For the emitter-follower circuit, find the following:

a. $z_{in(base)}$ c. v_{in} e. v_{out}
b. z_{in} d. A

Note: All the ac currents and voltages in this chapter will be in peak-to-peak values unless otherwise indicated.

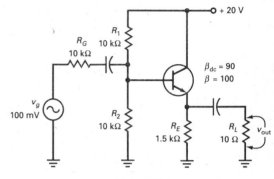

Figure 12-1

Steps	Comments
1. Draw the ac equivalent circuit. Figure 12-2	a. This ac equivalent circuit includes the model T. b. A preliminary analysis was done on this circuit and yielded the following results: $I_E = 6.2\text{ mA} \quad V_{CE} = 10.7\text{ V}$ $r'_e = 4.03\text{ }\Omega \quad R_1 // R_2 = 5\text{ k}\Omega$ $r_e = R_E // R_L = 9.93\text{ }\Omega$
2. Find $z_{in(base)}$. $z_{in(base)} = 100(9.93\text{ }\Omega + 4.03\text{ }\Omega) = 1.4\text{ k}\Omega$	Input impedance of the base. $z_{in(base)} = \beta(r_e + r'_e)$
3. Find z_{in}. $z_{in} = \dfrac{(5\text{ k}\Omega)(1.4\text{ k}\Omega)}{(5\text{ k}\Omega + 1.4\text{ k}\Omega)} = 1.09\text{ k}\Omega$	Input impedance of the stage. $z_{in} = R_1 // R_2 // \beta(r_e + r'_e)$
4. Find v_{in}. $v_{in} = \dfrac{1.09\text{ k}\Omega}{(1.09\text{ k}\Omega + 10\text{ k}\Omega)} 100\text{ mV} = 9.83\text{ mV}$	Voltage divider rule. $v_{in} = \dfrac{z_{in}}{z_{in} + R_G} v_g$
5. Find A. $A = \dfrac{9.93\text{ }\Omega}{9.93\text{ }\Omega + 4.03\text{ }\Omega} = 0.711$	Voltage gain of the stage (predicted). $A = \dfrac{r_e}{r_e + r'_e}$ always less than 1
6. Find v_{out}. $v_{out} = (0.711)(9.83\text{ mV}) = 6.99\text{ mV}$	From the gain formula. $v_{out} = Av_{in}$

Practice Problem 1. For the emitter-follower circuit, find the following:

a. $Z_{in(base)}$ c. v_{in} e. v_{out}
b. Z_{in} d. A

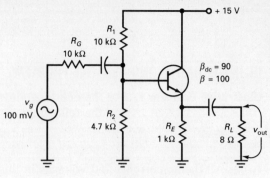

Figure 12-3

Answers: a. $Z_{in(base)} = 1.4 \text{ k}\Omega$ c. $v_{in} = 8.88 \text{ mV}$ e. $v_{out} = 5.03 \text{ mV}$
         b. $Z_{in} = 974 \ \Omega$ d. $A = 0.566$

Illustrative Problem 2. Find the maximum generator voltage that produces an unclipped output signal.

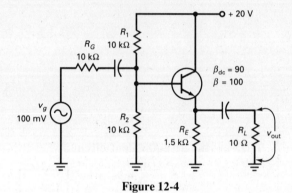

Figure 12-4

Steps	Comments
1. Draw the ac load line. (Optional) Figure 12-5	a. This is the same circuit as in Illustrative Problem 1, which yielded the following results: $I_E = 6.2 \text{ mA}$ $V_{CE} = 10.7 \text{ V}$ $r'_e = 4.03 \ \Omega$ $z_{in} = 1.09 \text{ k}\Omega$ $r_e = 9.93 \ \Omega$ $A = 0.711$ b. The ac load line is drawn to help illustrate how the MPP is found.
2. Find V_{CEQ}. $V_{CEQ} = 10.7 \text{ V}$	$V_{CE} = V_{CEQ}$
3. Find $I_{CQ}r_e$. $I_{CQ}r_e = (6.2 \text{ mA})(9.93 \ \Omega) = 61.6 \text{ mV}$	$I_E \approx I_{CQ}$
4. Find MPP. MPP = 2(61.6 mV) = 123 mV	a. If $V_{CEQ} < I_{CQ}r_e$, then MPP = $2V_{CEQ}$. b. If $V_{CEQ} > I_{CQ}r_e$, then MPP = $2I_{CQ}r_e$. Since $V_{CEQ} > I_{CQ}r_e$, therefore MPP = $2I_{CQ}r_e$.

Steps	Comments
5. Find v_{in}. (Maximum) $$v_{in} = \frac{123 \text{ mV}}{0.711} = 173 \text{ mV}$$	The maximum v_{in} for an unclipped output can be determined by using the value of MPP in the gain formula. $$v_{in} = \frac{v_{out}}{A} = \frac{MPP}{A}$$
6. Find v_g. (Maximum) $$v_g = \frac{10 \text{ k}\Omega + 1.09 \text{ k}\Omega}{1.09 \text{ k}\Omega} \, 173 \text{ mV} = 1.76 \text{ V}$$	Solving for v_g after applying the VDR. $$v_{in} = \frac{z_{in}}{R_g + z_{in}} v_g \Rightarrow v_g = \frac{R_g + z_{in}}{z_{in}} v_{in}$$

Practice Problem 2. Find the maximum generator voltage that produces an unclipped output signal. (This is the same circuit as in Practice Problem 1.)

Answer: $v_g = 1.3$ V

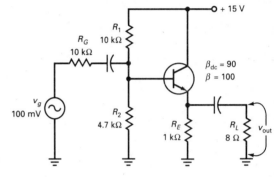

Figure 12-6

Illustrative Problem 3. For the Darlington amplifier, find the following:

a. $Z_{in(base)}$ **c.** V_{in} **e.** V_{out}
b. Z_{in} **d.** A

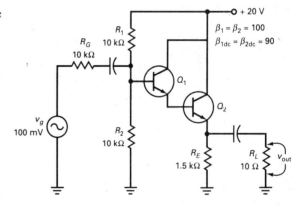

Figure 12-7

Steps	Comments
1. Find V_{TH}. $V_{TH} = \dfrac{10\ k\Omega}{10\ k\Omega + 10\ k\Omega}\ 20\ V = 10\ V$	The analysis of a Darlington amplifier is the same as an emitter follower except for the following: **a.** $\beta_{dc} = \beta_{1dc}\beta_{2dc} = (90)(90) = 8100$ **b.** The emitter current in Q_2 is
2. Find R_{TH}. $R_{TH} = \dfrac{(10\ k\Omega)(10\ k\Omega)}{10\ k\Omega + 10\ k\Omega} = 5\ k\Omega$	$I_E = \dfrac{V_{TH} - 1.4\ V}{R_E + R_{TH}/\beta_{dc}}$ $\approx \dfrac{V_{TH} - 1.4\ V}{R_E}$ (when $\beta_{dc}R_e > 20R_{TH}$)
3. Find I_E. $I_E = \dfrac{10\ V - 1.4\ V}{1.5\ k\Omega} = 5.73\ mA$	$\beta_{dc}R_E = 12.2\ M\Omega \qquad 20R_{TH} = 100\ k\Omega$ **c.** The total ac emitter resistance for the Darlington transistor is
4. Find r'_e. $r'_e = \dfrac{50\ mV}{5.73\ mA} = 8.73\ \Omega$	$r'_e = \dfrac{50\ mV}{I_E}$ (I_E is the emitter current in Q_2)
5. Find r_e. $r_e = 1.5\ k\Omega // 10\ \Omega = 9.93\ \Omega$	Total external ac emitter resistance. $r_e = R_E // R_L$
6. Find $z_{in(base)}$. $z_{in(base)} = 10\ K(9.93\ \Omega + 8.73\ \Omega) = 187\ k\Omega$	$\beta = \beta_1\beta_2 = (100)(100) = 10{,}000$ $z_{in(base)} = \beta(r_e + r'_e)$
7. Find z_{in}. $z_{in} = \dfrac{(5\ k\Omega)(187\ k\Omega)}{(5\ k\Omega + 187\ k\Omega)} = 4.87\ k\Omega$	Input impedance of the stage. $z_{in} = R_1 // R_2 // \beta(r_e + r'_e)$
8. Find v_{in}. $v_{in} = \dfrac{4.87\ k\Omega}{(4.87\ k\Omega + 10\ k\Omega)}\ 100\ mV = 32.8\ mV$	Voltage divider rule. $v_{in} = \dfrac{z_{in}}{z_{in} + R_G}\ v_g$
9. Find A. $A = \dfrac{9.93\ \Omega}{9.93\ \Omega + 8.73\ \Omega} = 0.532\ \Omega$	Voltage gain of the stage (predicted). $A = \dfrac{r_e}{r_e + r'_e}$ always less than 1
10. Find v_{out}. $v_{out} = (0.532)(32.8\ mV) = 17.4\ mV$	From the gain formula. $v_{out} = Av_{in}$

Practice Problem 3. For the Darlington amplifier, find the following:
- **a.** $z_{in(base)}$
- **b.** z_{in}
- **c.** v_{in}
- **d.** A
- **e.** v_{out}

Answers:
- **a.** $z_{in(base)} = 226\ k\Omega$
- **b.** $z_{in} = 3.15\ k\Omega$
- **c.** $v_{in} = 24\ mV$
- **d.** $A = 0.351$
- **e.** $v_{out} = 8.4\ mV$

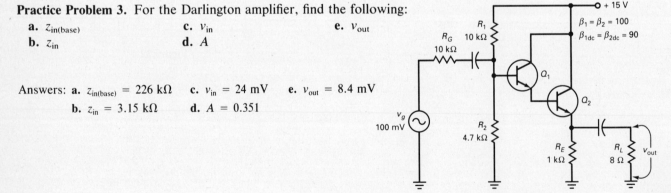

Figure 12-8

IV. PROBLEMS

Note: Assume $\beta_{dc} = 175$ and $\beta = 200$ in all problems of this chapter unless otherwise indicated.

Sec. 12-2   AC Model of a CC Amplifier

12-1. The circuit of Fig. 12-9 is an ac equivalent circuit of a CC amplifier with a T model transistor. The ac resistance of the emitter diode is 20 Ω. The total external emitter resistance is 100 Ω.
  a. Find $z_{in(base)}$.
  b. Find z_{in}.

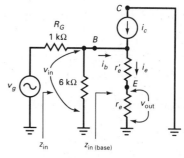

Figure 12-9

12-2. The circuit of Fig. 12-10 is an ac equivalent circuit of a CC amplifier with a Π model transistor. The ac resistance of the emitter diode is 12 Ω. The total external emitter resistance is 200 Ω.
  a. Find $z_{in(base)}$.
  b. Find z_{in}.

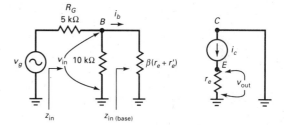

Figure 12-10

12-3. The circuit of Fig. 12-11 is an ac equivalent circuit of a CC amplifier. The ac resistance of the emitter diode is 10 Ω.
  a. Find $z_{in(base)}$.
  b. Find z_{in}.

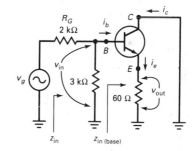

Figure 12-11

12-4. Find the input impedance of the CC amplifier in Fig. 12-12.

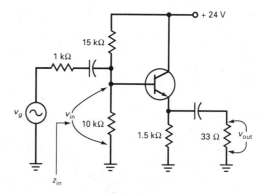

Figure 12-12

* 12-5. For the circuit of Fig. 12-13 and the data of circuit measurements, determine the input impedance of the CC amplifier.
  Data: $v_a = 1$ V $v_b = 400$ mV

Sec. 12-2 Voltage Gain

12-6. For the circuit of Fig. 12-13 and the data of circuit measurements, determine the voltage gain of the amplifier.
  Data: $v_b = 400$ mV $v_{out} = 360$ mV

12-7. With respect to Prob. 12-1, when $v_g = 100$ mV, find the following:
  a. v_{in} c. i_e e. A
  b. i_b d. v_{out} f. A (predicted)

12-8. With respect to Prob. 12-2, when $v_g = 200$ mV, find the following:
  a. v_{in} c. i_c e. A
  b. i_b d. v_{out} f. A (predicted)

12-9. What is the voltage gain of the CC amplifier in Fig. 12-11, when $r'_e = 15$ Ω?

12-10. For the circuit of Fig. 12-12, do the following:
  a. Find A.
  b. Find v_{out}, if $v_{in} = 500$ mV.

12-11. The dc emitter current is 10 mA in the circuit of Fig. 12-14. What is the voltage gain?

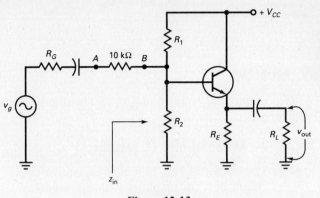

Figure 12-13

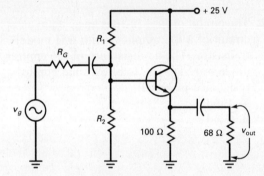

Figure 12-14

12-12. For the ac equivalent circuit of the CC amplifier in Fig. 12-15, do the following:
  a. Complete Table 12-1 for the given circuit conditions.

r'_e (Ω)	R_L (Ω)	β	z_{in} (kΩ)	A
5	22	200		
10	22	200		
5	100	200		
5	22	400		

Table 12-1

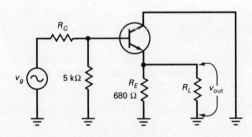

Figure 12-15

Select either *increased*, *decreased*, or *stayed the same* as the correct answer to the following questions.
  b. What happened to the input impedance when only r'_e was increased?
  c. What happened to the input impedance when only R_L was increased?
  d. What happened to the input impedance when only β was increased?
  e. What happened to the voltage gain when only r'_e was increased?
  f. What happened to the voltage gain when only R_L was increased?
  g. What happened to the voltage gain when only β was increased?

* See "Practical Techniques," Chap. 10.

12-13. Find the output voltage of the emitter follower in Fig. 12-16.

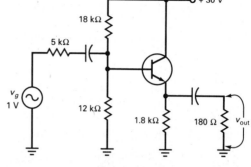

Figure 12-16

12-14. Find the output voltage of the emitter follower in Fig. 12-17.

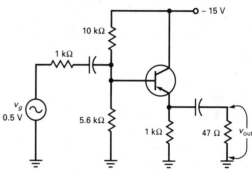

Figure 12-17

Sec. 12-4 Maximum Unclipped Output

12-15. For the circuit of Fig. 12-18, do the following:
  a. Find the MPP.
  b. Find the maximum generator voltage that produces an unclipped output signal.

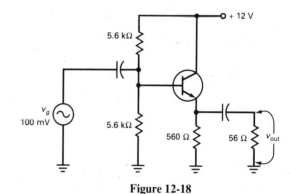

Figure 12-18

12-16. For the circuit of Fig. 12-16, do the following:
  a. Find the MPP.
  b. Find the maximum generator voltage that produces an unclipped output signal.
  c. Find the MPP if R_L is decreased to 100 Ω.

12-17. For the circuit of Fig. 12-17, do the following:
  a. Find the MPP.
  b. Find the maximum generator voltage that produces an unclipped output signal.
  c. Find the MPP if R_L is increased to 100 Ω.

Sec. 12-5 Cascading CE and CC

12-18. For the circuit of Fig. 12-19, find the following:
  a. z_{in} (first stage)
  b. $z_{in(base)}$ (second stage)
  c. A (first stage)
  d. A (second stage)
  e. A (total gain)
  f. v_{out}

12-19. What is the ac output voltage of the second stage in the circuit of Fig. 12-19 if $\beta = 400$ for both transistors?

12-20. For the circuit of Fig. 12-19, do the following:
  a. Find the MPP (first stage).
  b. Find the MPP (second stage).
  c. Find the maximum generator voltage that produces an unclipped signal at the load.
  (*Hint:* Calculate the maximum generator voltage for each MPP; your answer will be the smaller of the two.)

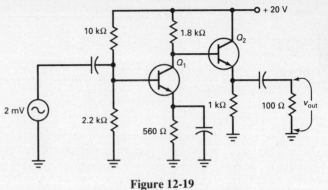

Figure 12-19

12-21. For the circuit of Fig. 12-20, find the following:
  a. z_{in} (first stage)
  b. z_{in} (second stage)
  c. A (first stage)
  d. A (second stage)
  e. A (total gain)
  f. v_{out}

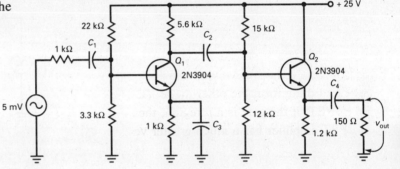

Figure 12-20

12-22. For the circuit of Fig. 12-20, do the following:
  a. Find the MPP (first stage).
  b. Find the MPP (second stage).
  c. Find the maximum generator voltage that produces an unclipped signal at the load.

Sec. 12-6 Darlington Transistor

12-23. For the Darlington amplifier in Fig. 12-21, find the following:
  a. I_E (second transistor)
  b. I_E (first transistor)
  c. V_{CE} (second transistor)
  d. V_{CE} (first transistor)
  e. r'_e (Darlington transistor)

12-24. For the Darlington amplifier in Fig. 12-21, find the following:
  a. $z_{in(base)}$
  b. z_{in}
  c. v_{in}
  d. A
  e. v_{out}

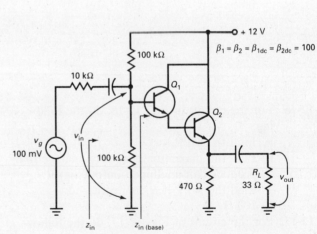

Figure 12-21

12-25. What is the output voltage in the circuit of Fig. 12-21 if β is 200 for both transistors?

12-26. What is the output voltage in the circuit of Fig. 12-21 if R_L is 10 Ω?

CHAPTER 13
FIELD-EFFECT TRANSISTORS

Study Chap. 13 in *Electronic Principles*.

I. TRUE / FALSE

Answer true (T) or false (F) to each of the following statements.

Answer

1. The JFET is similar to two diodes; the gate and source form one diode and the gate and drain form the other diode. (13-1)
   1. ___
2. In a JFET, if the gate is a *p* region, then the source and drain are *n* regions. (13-1)
   2. ___
3. A JFET amplifier has a much greater voltage gain than a bipolar transistor. (13-2)
   3. ___
4. The source and drain are not interchangeable in many low-frequency applications. (13-2)
   4. ___
5. The gate current of a JFET is approximately zero. (13-2)
   5. ___
6. The term field effect is related to how the gate voltage can control the current through the channel. (13-2)
   6. ___
7. A typical JFET has an input resistance in the hundreds of kilohms. (13-2)
   7. ___
8. The maximum drain current out of a JFET occurs when the gate-source voltage is zero. (13-3)
   8. ___
9. The magnitudes of $V_{GS(off)}$ and V_P are equal. (13-3)
   9. ___
10. The saturation region of a bipolar transistor is equivalent to the ohmic region of a JFET. (13-3)
    10. ___
11. V_{DS} is the value of the pinchoff voltage where the highest drain current changes from almost a vertical line to almost a horizontal line on the drain curves. (13-3)
    11. ___
12. JFETs are often called square-law devices. (13-4)
    12. ___
13. The drain current of a JFET can be calculated if the values of I_{DSS}, $V_{GS(off)}$, and V_{GS} are known.
    13. ___
14. When a JFET is operating in the ohmic region of the drain curves, it acts like a resistor. (13-5)
    14. ___
15. When a JFET is operating on the horizontal part of the drain curves, it acts as a constant voltage source. (13-5)
    15. ___
16. The drain curves of an ideal JFET do not have a breakdown region. (13-5)
    16. ___
17. A MOSFET has a gate that is electrically insulated from the channel. (13-6)
    17. ___
18. The depletion-mode MOSFET can operate when either a positive or negative voltage is applied to the gate. (13-6)
    18. ___

19. The maximum drain current cannot exceed the value of I_{DSS} for a depletion-mode MOSFET. (13-6)   19. ___
20. In an enhancement MOSFET, a conducting layer between the source and the drain is created when V_{GS} is made greater than $V_{GS(th)}$. (13-7)   20. ___
21. Use an ohmic equivalent circuit for an enhancement-mode MOSFET, when V_{DS} is greater than V'_K. (13-7)   21. ___
22. FET data sheets are similar to bipolar data sheets. (13-8)   22. ___

II. COMPLETION

Complete each of the following statements.   Answer

1. The gate of a JFET is analogous to the ___ of a bipolar transistor. (13-1)   1. _____
2. The drain of a JFET is analogous to the ___ of a bipolar transistor. (13-1)   2. _____
3. The gate-source diode of a JFET is always ___-bias. (13-2)   3. _____
4. A JFET acts as a ___-control device. (13-2)   4. _____
5. The source current of a JFET is equal to the ___ current. (13-2)   5. _____
6. The drain and source are p-type semiconductor material in a ___-channel JFET. (13-2)   6. _____
7. The gate-to-source voltage of a JFET controls the ___ of the channel. (13-2)   7. _____
8. The maximum voltage $V_{DS(max)}$ is called the ___ voltage. (13-3)   8. _____
9. The minimum voltage that V_{DS} can be in order that the JFET acts like a constant current source with a value of I_{DSS} is called the ___ voltage. (13-3)   9. _____
10. The almost vertical part of the drain curves is called the ___ region. (13-3)   10. _____
11. When V_{GS} is equal to $V_{GS(off)}$, the drain current is equal to ___. (13-3)   11. _____
12. The transconductance curve of a JFET is a graph of I_D versus ___. (13-4)   12. _____
13. The geometric shape of the transconductance curve is part of a ___. (13-4)   13. _____
14. The value of R_{DS} is equal to V_P divided by ___. (13-5)   14. _____
15. The voltage that borders the ohmic region and the constant source region for any value of V_{GS} other than zero is called the ___ pinchoff voltage. (13-5)   15. _____
16. In order for the JFET to operate as a constant current source, V_{DS} must be greater than ___. (13-5)   16. _____
17. The MOSFET is often referred to as an ___. (13-6)   17. _____
18. The body of a depletion-mode MOSFET is called the ___. (13-6)   18. _____
19. The drain current for a depletion-mode MOSFET with a shorted gate-source voltage is equal to ___. (13-6)   19. _____
20. The minimum V_{GS} that creates the n-type inversion layer in an enhancement-mode MOSFET is called the ___ voltage. (13-7)   20. _____
21. When V_{GS} is zero in an enhancement-mode MOSFET, the drain current is ___. (13-7)   21. _____
22. The voltage that borders the ohmic region and the constant current source region of an ideal enhancement-mode MOSFET is called the proportional ___ voltage. (13-7)   22. _____
23. Two of the most important pieces of information on a data sheet of a depletion-mode device are the value of I_{DSS} and ___. (13-8)   23. _____
24. Enhancement-mode devices have three key quantities, which are $I_{D(on)}$, $V_{GS(on)}$, and ___. (13-8)   24. _____

III. ILLUSTRATIVE AND PRACTICE PROBLEMS

Illustrative Problem 1. For the given circuit find the following:
a. I_D
b. V_{DS}

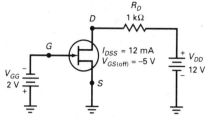

Figure 13-1

Steps	Comments
1. Find V_{GS}. $\quad V_{GS} = -2\text{ V} - 0\text{ V} = -2\text{ V}$	$V_G = -2\text{ V} \quad V_S = 0\text{ V}$ $V_{GS} = V_G - V_S$
2. Find V_P. $\quad V_P = 5\text{ V}$	Pinchoff voltage. $V_P = -V_{GS(off)}$
3. Find I_D. $\quad I_D = 12\text{ mA}\left(1 - \dfrac{-2\text{ V}}{-5\text{ V}}\right)^2 = 4.32\text{ mA}$	Assume that the JFET is operating in the constant current source region. ($V_{DS} > V'_P$ is a condition that must be met for the assumption to be valid.) $I_D = I_{DSS}\left(1 - \dfrac{V_{GS}}{V_{GS(off)}}\right)^2$
4. Find V_{DS}. Figure 13-2 $V_{DS} = 12\text{ V} - (4.32\text{ mA})(1\text{ k}\Omega) = 7.68\text{ V}$	a. Use the constant current source equivalent drain circuit. $V_{DS} = V_{DD} - I_D R_D$ b. If $V_{DS} \geq V_P$ (which it is), then V_{DS} must be greater than V'_P, since $V_P \geq V'_P$ for a JFET. c. The assumption is correct. Answers stand.

Practice Problem 1. For the given circuit find the following:
a. I_D
b. V_{DS}

Answers: a. $I_D = 5\text{ mA}$ b. $V_{DS} = 9\text{ V}$

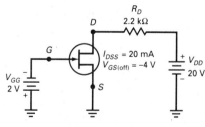

Figure 13-3

Illustrative Problem 2. For the given circuit find the following:
a. I_D
b. V_{DS}

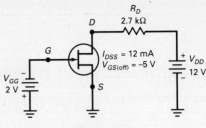

Figure 13-4

Steps	Comments
1. Find V_{GS}. $V_{GS} = -2$ V	Same as Illustrative Problem 1.
2. Find V_P. $V_P = 5$ V	Same as Illustrative Problem 1.
3. Find I_D. $I_D = 4.32$ mA	Same as Illustrative Problem 1.
4. Find V_{DS}. $V_{DS} = 12\text{ V} - (4.32\text{ mA})(2.7\text{ k}\Omega) = 0.336\text{ V}$	a. Same as Illustrative Problem 1. b. Since V_{DS} is now $< V_P$, check to see if $V_{DS} < V'_P$.
5. Find R_{DS}. $R_{DS} = \dfrac{5\text{ V}}{12\text{ mA}} = 0.417\text{ k}\Omega$	$R_{DS} = \dfrac{V_P}{I_{DSS}}$
6. Find V'_P (for $I_D = 4.32$ mA). $V'_P = (4.32\text{ mA})(0.417\text{ k}\Omega) = 1.8\text{ V}$	a. Proportional pinchoff voltage. $V'_P = I_D R_{DS}$ b. Since $V_{DS} < V'_P$, the initial assumption is incorrect. Therefore, the JFET is operating in the ohmic region.
7. Find I_D and V_{DS}. [Figure 13-5: circuit with R_{DS} 417 Ω, R_D 2.7 kΩ, V_{DD} 12 V] Figure 13-5 $I_D = \dfrac{12\text{ V}}{2.7\text{ k}\Omega + 0.417\text{ k}\Omega} = 3.85\text{ mA}$ $V_{DS} = (3.85\text{ mA})(0.417\text{ k}\Omega) = 1.61\text{ V}$	a. Disregard the previous values of I_D and V_{DS}. b. Now use the ohmic equivalent drain circuit to find I_D and V_{DS}. $I_D = \dfrac{V_{DD}}{R_D + R_{DS}}$ $V_{DS} = I_D R_{DS}$ c. By the way, in the ohmic region V'_P is equal to V_{DS}. $V'_P = 1.61\text{ V}$ for $I_D = 3.85\text{ mA}$

Practice Problem 2. For the given circuit find the following:
a. I_D
b. V_{DS}

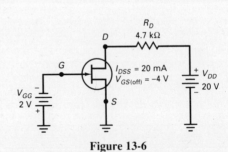

Figure 13-6

Answers: a. $I_D = 4.08$ mA   b. $V_{DS} = 0.816$ V

Illustrative Problem 3. For the given circuit find the following:
a. I_D
b. V_{DS}

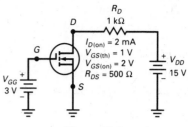

Figure 13-7

Steps	Comments
1. Find V_{GS}. $V_{GS} = 3\text{ V} - 0\text{ V} = 3\text{ V}$	$V_G = 3\text{ V} \quad V_S = 0\text{ V} \quad V_{GS} = V_G - V_S$
2. Find K. $K = \left(\dfrac{3\text{ V} - 1\text{ V}}{2\text{ V} - 1\text{ V}}\right)^2 = 4$	$K = \left(\dfrac{V_{GS} - V_{GS(th)}}{V_{GS(on)} - V_{GS(th)}}\right)^2$
3. Find I_D. $I_D = 4(2\text{ mA}) = 8\text{ mA}$	Assume that the JFET is operating in the constant current source region. ($V_{DS} \geq V'_K$ is a condition that must be met for the assumption to be valid.) $I_D = K I_{D(on)}$
4. Find V'_K. $V'_K = (8\text{ mA})(0.5\text{ k}\Omega) = 4\text{ V}$	Proportional knee voltage. $V'_K = I_D R_{DS}$
5. Find V_{DS}. *(Figure 13-8: constant current source equivalent drain circuit with $R_D = 1\text{ k}\Omega$, $V_{DD} = 15\text{ V}$)* $V_{DS} = 15\text{ V} - (8\text{ mA})(1\text{ k}\Omega) = 7\text{ V}$	a. Use the constant current source equivalent drain circuit. $V_{DS} = V_{DD} - I_D R_D$ b. Compare V_{DS} to V'_K. c. If $V_{DS} > V'_K$ (which it is), then the assumption is correct. Answers stand. d. However, if $V_{DS} < V'_K$, you would have to use the ohmic equivalent drain circuit to find I_D and V_{DS}.

Practice Problem 3. For the given circuit find the following:
a. I_D
b. V_{DS}

Answers: a. $I_D = 2.22\text{ mA}$ b. $V_{DS} = 10.7\text{ V}$

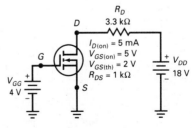

Figure 13-9

IV. PROBLEMS

Sec. 13-2 The Biased JFET

13-1. The gate circuit is 10 nA in the circuit of Fig. 13-10. What is the dc input resistance of the gate?

13-2. What is the minimum dc input resistance of a JFET whose gate current can vary from 2 nA to 20 nA when $V_{GS} = -10$ V?

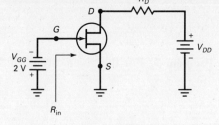

Figure 13-10

Sec. 13-3 Drain Curves

13-3. For the JFET's drain curves that are shown in Fig. 13-11, find the following:
  a. I_{DSS}
  b. $V_{GS(off)}$
  c. V_P
  d. $V_{DS(max)}$ (at $V_{GS(off)}$)
  e. R_{DS}

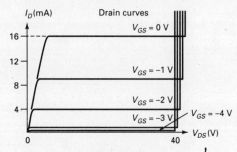

Figure 13-11

13-4. A JFET has $I_{DSS} = 15$ mA and $V_{GS(off)} = -5$ V.
  a. Find the maximum drain current.
  b. Find the minimum drain current.
  c. Find V_P.
  d. Find R_{DS}.

13-5. A p-channel JFET has $R_{DS} = 500$ Ω and $I_{DSS} = 10$ mA. What does $V_{GS(off)}$ equal?

Sec. 13-4 The Transconductance Curve

13-6. A JFET has $I_{DSS} = 20$ mA and $V_{GS(off)} = -5$ V.
  a. Complete Table 13-1 for the given values of V_{GS}.
  b. From the values of Table 13-1, plot the transconductance curve for the JFET on engineering graph paper.

13-7. For the JFET's drain curves that are shown in Fig. 13-11, plot its transconductance curve on engineering graph paper.

13-8. The transconductance curve of a JFET is drawn along side its drain curves in Fig. 13-12. Fill in the missing values of I_D and V_{GS} between the appropriate parentheses.

V_{GS} V	I_D mA
0	
−1	
−2	
−3	
−4	
−5	

Table 13-1

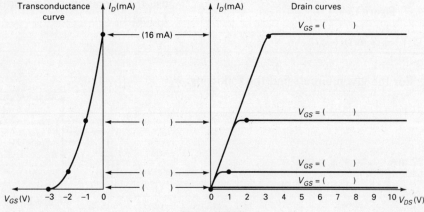

Figure 13-12

13-9. For the circuit of Fig. 13-13, when $R_D = 2.2\ k\Omega$, find the following:
a. V_{GS} d. I_D
b. R_{DS} e. V'_P
c. V_P f. V_{DS}

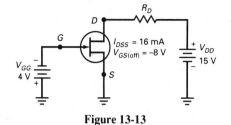

Figure 13-13

13-10. For the circuit of Fig. 13-13, when $R_D = 5.6\ k\Omega$, find the following:
a. I_D
b. V_{DS}

13-11. For the circuit of Fig. 13-14, find the following:
a. I_D
b. V_{DS}

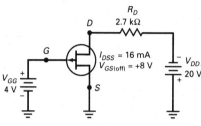

Figure 13-14

13-12 For the circuit of Fig. 13-15, do the following:
a. Complete Table 13-2 for the given circuit conditions.

V_{GS} V	V_{DD} V	R_D kΩ	I_D mA	V_{DS} V
−2	12	1		
−1	12	1		
−2	20	1		
−2	12	2.2		

Table 13-2

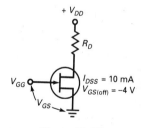

Figure 13-15

Select either *increased*, *decreased*, or *stayed the same* as the correct answer to the following questions.
b. What happened to I_D when only V_{GS} was increased?
c. What happened to V_{DS} when only V_{GS} was increased?
d. What happened to I_D when only V_{DD} was increased?
e. What happened to V_{DS} when only V_{DD} was increased?
f. What happened to I_D when only R_D was increased?
g. What happened to V_{DS} when only R_D was increased?

13-13. The same JFET is subjected to the circuit conditions that are shown in Fig. 13-16. From the voltmeter readings, determine I_{DSS} and the approximate $V_{GS(off)}$. *Note:* When I_D approaches zero in the constant current mode, V_{GS} approaches $V_{GS(off)}$.

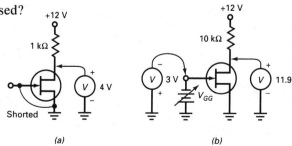

Figure 13-16

Sec. 13-6 The Depletion-Mode MOSFET

13-14. For the circuit of Fig. 13-17, when $R_D = 1.5\ k\Omega$, find the following:
- a. V_{GS}
- b. R_{DS}
- c. V_P
- d. I_D
- e. V'_P
- f. V_{DS}

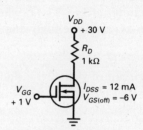

Figure 13-17

13-15. For the circuit of Fig. 13-17, when $R_D = 5.6\ k\Omega$, find the following:
- a. I_D
- b. V_{DS}

13-16. For the circuit of Fig. 13-18, find the following:
- a. I_D
- b. V_P
- c. V'_P
- d. V_{DS}

13-17. For the circuit of Fig. 13-19, do the following:
- a. Complete Table 13-3 for the given circuit conditions.

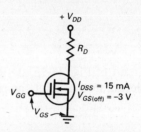

Figure 13-18

V_{GS} V	V_{DD} V	R_D kΩ	I_D mA	V_{DS} V
−1	20	1		
0	20	1		
−1	30	1		
−1	20	2.2		

Table 13-3

Figure 13-19

Select either *increased*, *decreased*, or *stayed the same* as the correct answer to the following questions.
- b. What happened to I_D when only V_{GS} was increased?
- c. What happened to V_{DS} when only V_{GS} was increased?
- d. What happened to I_D when only V_{DD} was increased?
- e. What happened to V_{DS} when only V_{DD} was increased?
- f. What happened to I_D when only R_D was increased?
- g. What happened to V_{DS} when only R_D was increased?

13-18. For the circuit of Fig. 13-20, find the following:
- a. I_D
- b. V_{DS}

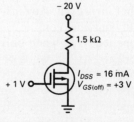

Figure 13-20

Sec. 13-7 The Enhancement-Mode MOSFET

13-19. The equation that generates the transconductance curve of the enhancement-mode MOSFET is $I_D = k(V_{GS} - V_{GS(th)})^2$. Calculate the value of k for the transconductance curve of Fig. 13-21.

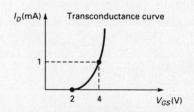

Figure 13-21

13-20. An enhancement-mode MOSFET has $I_{D(on)} = 8$ mA, $V_{GS(on)} = 5$ V, and $V_{GS(th)} = 1$ V.
  a. Calculate the value of k.
  b. Complete Table 13-4 for the given values of V_{GS}.
  c. From the values of Table 13-4, plot the transconductance curve for the MOSFET on engineering graph paper.

13-21. A variation of the equation in Prob. 13-19 is
$I_D = KI_{D(on)}$ where

$$K = \left(\frac{V_{GS} - V_{GS(th)}}{V_{GS(on)} - V_{GS(th)}}\right)^2$$

  a. Calculate the value of K for the circuit conditions that are shown in Fig. 13-22. (Assume that $V_{DS} > V'_K$.)
  b. Find I_D.
  c. Find V_{DS}.
  d. Find V'_K.

V_{GS} V	I_D mA
1	
2	
3	
4	
5	
6	

Table 13-4

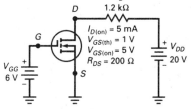

Figure 13-22

13-22. For the circuit of Fig. 13-23, when $R_D = 1.5$ kΩ, find the following:
  a. V_{GS}
  b. I_D
  d. V_{DS}
  e. V'_K

13-23. For the circuit of Fig. 13-23, when $R_D = 4.7$ kΩ, find the following:
  a. I_D
  b. V_{DS}

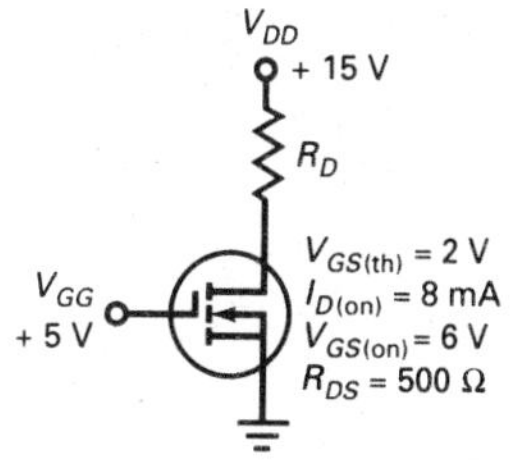

13-24. For the circuit of Fig. 13-24, find the following:
  a. I_D
  b. V_{DS}

13-25. For the circuit of Fig. 13-25, do the following:
  a. Complete Table 13-5 for the given circuit conditions.

Figure 13-23

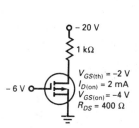

Figure 13-24

V_{GS} V	V_{DD} V	R_D kΩ	I_D mA	V_{DS} V	V'_K V
7	16	1.5			
8	16	1.5			
7	24	1.5			
7	16	2.7			

Table 13-5

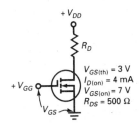

Figure 13-25

Select either *increased*, *decreased*, or *stayed the same* as the correct answer to the following questions.
  b. What happened to I_D when only V_{GS} was increased?
  c. What happened to V_{DS} when only V_{GS} was increased?

    d. What happened to I_D when only V_{DD} was increased?
    e. What happened to V_{DS} when only V_{DD} was increased?
    f. What happened to I_D when only R_D was increased?
    g. What happened to V_{DS} when only R_D was increased?

Sec. 13-8 Reading Data Sheets

13-26. What is the minimum dc input resistance of the MPF102 JFET at room temperature for the conditions that are specified on its data sheet?

13-27. Would a MPF102 JFET be operating within the manufacturer's specifications if its drain current were 30 mA when $V_{DS} = 15$ V and $V_{GS} = 0$ V?

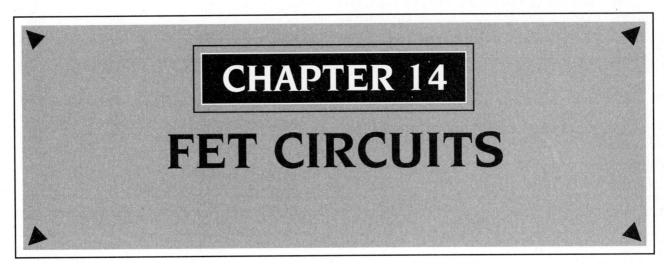

CHAPTER 14
FET CIRCUITS

Study Chap. 14 in *Electronic Principles*.

I. TRUE / FALSE

Answer true (T) or false (F) to each of the following statements.

Answer

1. A JFET self-bias circuit has no equivalent bipolar bias circuit. **(14-1)** 1. ___
2. A bipolar voltage-divider bias circuit has an equivalent JFET bias circuit. **(14-1)** 2. ___
3. BJT's emitter-feedback and collector-feedback bias circuits have no equivalent JFET bias circuits. **(14-1)** 3. ___
4. Positive feedback is utilized in the JFET self-bias circuit operation. **(14-1)** 4. ___
5. When R_S is large in a self-bias JFET circuit, the Q point is far up the transconductance curve. **(14-2)** 5. ___
6. The optimum Q point is when it is located near the middle of the transconductance curve. **(14-2)** 6. ___
7. The graphical solution for the self-bias JFET assumes that the JFET is operating in the constant current mode. **(14-2)** 7. ___
8. The variable on the vertical axis of the universal curve of a self-bias JFET is R_S/R_{DS}. **(14-3)** 8. ___
9. The universal curve can be used for any self-biased JFET. **(14-3)** 9. ___
10. The transconductance of a JFET is equal to a change in I_D divided by a corresponding change in V_{GS}. **(14-4)** 10. ___
11. The plot of g_m versus V_{GS} for a JFET is nonlinear. **(14-4)** 11. ___
12. The source follower is one of the most widely used JFET circuits. **(14-5)** 12. ___
13. A JFET used as an amplifier should act as a current source, not as a resistor. **(14-5)** 13. ___
14. The on-off ratio of a JFET shunt switch is much higher than a JFET series switch with the same R_D. **(14-6)** 14. ___
15. When the JFET switch is closed, the JFET is equivalent to a resistance that is equal R_{DS}. **(14-6)** 15. ___
16. The depletion-mode MOSFET amplifier can be uniquely biased so that V_{GS} is equal to zero. **(14-7)** 16. ___
17. The JFET formulas for the voltage gain apply directly to a MOSFET depletion-mode amplifier. **(14-7)** 17. ___
18. The enhancement-mode MOSFET has revolutionized the computer industry in its application as a general-purpose amplifier. **(14-8)** 18. ___

19. The VMOS transistor cannot be used in high-speed switching circuits because it takes to much time to come out of saturation. (14-8)   19. _____
20. The VMOS is an enhancement-mode MOSFET modified to handle much larger currents and voltages than a conventional MOSFET. (14-8)   20. _____
21. The circuit that provides the ultimate bias stability for a JFET is current-source bias. (14-9)   21. _____

II. COMPLETION

Complete each of the following statements.

Answer

1. A BJT two-supply emitter bias is equivalent to a JFET two-supply _____ bias. (14-1)
2. The basic idea behind self-bias is to use the voltage across _____ to produce the required gate-source bias voltage. (14-1)
3. The least preferred form of biasing for a JFET amplifier is through _____ bias. (14-1)
4. The bias circuit of a JFET should always cause the gate-source diode to be _____-bias. (14-1)
5. The minimum number of points required to draw the self-bias line on the transconductance curve is _____. (14-2)
6. One of the points of the self-bias line will always be located at the _____ of the transconductance curve coordinates. (14-2)
7. The coordinate values of the point of interception between the self-bias line and the transconductance curve provide the _____ values of I_D and V_{GS}. (14-2)
8. The maximum value for the variable on the vertical axis of the universal JFET curve is _____. (14-3)
9. The quantities that are needed to solve for I_D using the universal self-bias JFET curve are I_{DSS}, $V_{GS(off)}$, and _____. (14-3)
10. The variable on the horizontal axis of the universal curve is _____. (14-3)
11. The formal equivalent for the mho is the _____. (14-4)
12. The transconductance of a JFET reaches its maximum value when V_{GS} is equal to _____. (14-4)
13. The voltage gain of a fully by-passed common-source amplifier is equal to r_d times _____. (14-5)
14. The voltage gain of a source follower is always less than _____. (14-5)
15. The voltage that turns the JFET analog switch on and off is _____. (14-6)
16. When a JFET is being used as an analog switch, it is operating in its _____ region. (14-6)
17. The g_m of a MOSFET can be controlled by changing the _____-_____ voltage. (14-7)
18. The quality of the low-noise properties in a MOSFET is _____. (14-7)
19. Since the size of components is sometimes crucial in integrated circuit design, a MOSFET can be also used as an _____-_____ resistor. (14-8)
20. A VMOS transistor has a negative thermal coefficient, which means that as the device's temperature increases, its drain current _____. (14-8)
21. The stability of a JFET voltage-divider bias circuit depends on how much V_{TH} is greater than _____. (14-9)

III. ILLUSTRATIVE AND PRACTICE PROBLEMS

Illustrative Problem 1. For the given circuit use the graphical method to find the following:

a. I_D
b. V_{GS}
c. V_{DS}

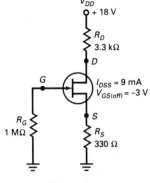

Figure 14-1

Steps	Comments
1. Draw the self-bias line on the plotted transconductance curve to determine I_D and V_{GS}. 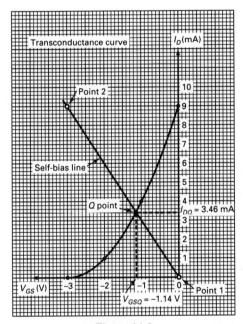 Figure 14-2 $I_D = 3.46$ mA $V_{GS} = -1.14$ V	a. Assume the JFET is operating as a constant current source. b. Plot the transconductance curve using the following equation: $$I_D = I_{DSS}\left(1 - \frac{V_{GS}}{V_{GS(off)}}\right)^2$$ | V_{GS}	0	-1 V	-2 V	-3 V | |---|---|---|---|---| | I_D | 9 mA | 4 mA | 1 mA | 0 | Table 14-1 c. The self-bias line equation [$V_{GS} = -I_D(330\ \Omega)$] is derived from circuit conditions. Points that exit on the self-bias line can be determined by substituting values of I_D into this equation. Point no. 1: When $I_D = 0$, then $V_{GS} = 0$. Point no. 2: When $I_D = 9$ mA, then $V_{GS} = -2.97$ V. Draw the self-bias line through these two points. d. The quiescent values for I_D and V_{GS} are the coordinates of the point of interception.
2. Find V_{DS}. $V_{DS} = 18$ V $- 3.46$ mA$(3.3$ k$\Omega + 0.33$ k$\Omega) = 5.44$ V	$V_{DS} = V_{DD} - I_D(R_D + R_S)$ Since $V_{DS} > V_P$, the initial assumption is correct. ($V_P = -V_{GS(off)} = 3$ V)

Practice Problem 1. For the given circuit use the graphical method to find the following:

a. I_D
b. V_{GS}
c. V_{DS}

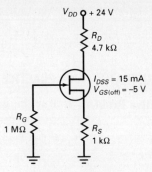

Answers: **a.** $I_D = 2.83$ mA **b.** $V_{GS} = -2.83$ V **c.** $V_{DS} = 7.87$ V

Figure 14-3

Illustrative Problem 2. For the given common source amplifier, find the following:

a. g_{m0}
b. g_m
c. z_{in}
d. v_{in}
e. A
f. v_{out}

Note: All ac currents and voltages will be in peak-to-peak values, unless otherwise indicated.

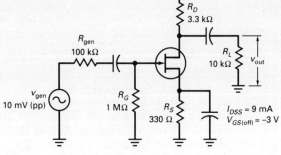

Figure 14-4

Steps	Comments
1. Draw the ac equivalent circuit. Figure 14-5	a. The dc biasing conditions are the same as in Illustrative Problem 1. $I_D = 3.46$ mA $V_{GS} = -1.14$ V $V_{DS} = 5.44$ V b. Since the gate of a JFET has a extremely high impedance, you can consider it to be an opened circuit. c. $i_d = g_m v_{gs}$, $v_{in} = v_g = v_{gs}$ d. $r_d = R_L R_D/(R_L + R_D) = 2.48$ kΩ
2. Find g_{m0}. $g_{m0} = \dfrac{2(9 \text{ mA})}{3 \text{ V}} = 6000$ μmho	Slope of the transconductance curve where V_{GS} is equal to zero. $g_{m0} = \dfrac{2 I_{DSS}}{V_P}$
3. Find g_m. $g_m = 6000$ μmho $\sqrt{\dfrac{3.46 \text{ mA}}{9 \text{ mA}}} = 3720$ μmho	Slope of the transconductance curve at the Q point. $g_m = g_{m0} \sqrt{\dfrac{I_D}{I_{DSS}}}$
4. Find z_{in}. $z_{in} = 1$ MΩ	Input impedance of the stage. $z_{in} = R_G$
5. Find v_{in}. $v_{in} = \dfrac{1 \text{ MΩ}}{1 \text{ MΩ} + 100 \text{ kΩ}} 10 \text{ mV} = 9.09$ mV	Voltage divider rule. $v_{in} = \dfrac{R_G}{R_G + R_{gen}} v_{gen}$

Steps	Comments
6. Find A. $A = 3720 \ \mu\text{mho}(2.48 \ \text{k}\Omega) = 9.23$	Voltage gain of the stage. $A = g_m r_d$
7. Find v_{out}. $v_{\text{out}} = 9.23(9.09 \ \text{mV}) = 83.9 \ \text{mV}$	Output voltage of the stage. $v_{\text{out}} = A v_{\text{in}}$

Practice Problem 2. For the given common source amplifier, find the following:

a. g_{m0} d. v_{in}
b. g_m e. A
c. Z_{in} f. v_{out}

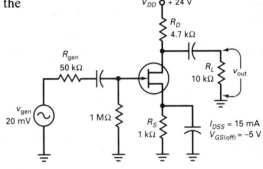

Answers: a. $g_{m0} = 6000 \ \mu\text{mho}$ d. $v_{\text{in}} = 19 \ \text{mV}$
b. $g_m = 2600 \ \text{mho}$ e. $A = 8.33$
c. $Z_{\text{in}} = 1 \ \text{M}\Omega$ f. $v_{\text{out}} = 159 \ \text{mV}$

Figure 14-6

Illustrative Problem 3. For the given source follower with a $g_m = 4080 \ \mu\text{mho}$, do the following:

a. Find v_{in}.
b. Find A.
c. Find v_{out}.
d. Find i_d by $i_d = g_m v_{gs}$.
e. Find i_d by $i_d = v_{\text{out}}/r_s$.

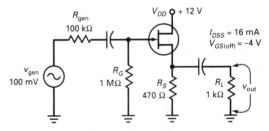

Figure 14-7

Steps	Comments
1. Draw the ac equivalent circuit. Figure 14-8	a. The dc analysis that was done on this circuit and yielded the following results: $I_D = 4.17 \ \text{mA}$ $V_{GS} = -1.96 \ \text{V}$ $V_{DS} = 10 \ \text{V}$ $V_P = 4 \ \text{V}$ $g_{m0} = 8000 \ \mu\text{mho}$ $g_m = 4080 \ \mu\text{mho}$ b. $r_s = R_S R_L / (R_S + R_L) = 320 \ \Omega$ c. $Z_{\text{in}} = R_G = 1 \ \text{M}\Omega$ d. $v_{\text{out}} = v_s, \ v_{\text{in}} = v_g$
2. Find v_{in}. $v_{\text{in}} = \dfrac{1 \ \text{M}\Omega}{1 \ \text{M}\Omega + 100 \ \text{k}\Omega} 100 \ \text{mV} = 90.9 \ \text{mV}$	Voltage divider rule. $v_{\text{in}} = \dfrac{R_G}{R_G + R_{\text{gen}}} v_{\text{gen}}$

Steps	Comments
3. Find A. $A = \dfrac{320\ \Omega}{320\ \Omega + 1/4080\ \mu\text{mho}} = 0.566$	Voltage gain of the stage. $A = \dfrac{r_s}{r_s + 1/g_m}$
4. Find v_{out}. $v_{\text{out}} = 0.566(90.9\ \text{mV}) = 51.5\ \text{mV}$	Output voltage of the stage. $v_{\text{out}} = A v_{\text{in}}$
5. Find i_d. ($i_d = g_m v_{gs}$) $i_d = (90.9\ \text{mV} - 51.5\ \text{mV})4080\ \mu\text{mho} = 161\ \mu\text{A}$	$v_{gs} = v_g - v_s = v_{\text{in}} - v_{\text{out}}$ $i_d = v_{gs} g_m = (v_{\text{in}} - v_{\text{out}}) g_m$
6. Find i_d. ($i_d = v_{\text{out}}/r_s$) $i_d = 51.5\ \text{mV}/320\ \Omega = 161\ \mu\text{A}$	Each method should provide the same value of the ac drain current.

Practice Problem 3. For the given source follower with a $g_m = 5120\ \mu\text{mho}$, do the following:
a. Find v_{in}.
b. Find A.
c. Find v_{out}.
d. Find i_d by $i_d = g_m v_{gs}$.
e. Find i_d by $i_d = v_{\text{out}}/r_s$.

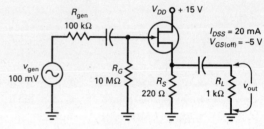

Figure 14-9

Answers: a. $v_{\text{in}} = 99\ \text{mV}$ d. $i_d = 264\ \mu\text{A}$
b. $A = 0.48$ e. $i_d = 264\ \mu\text{A}$
c. $v_{\text{out}} = 47.5\ \text{mV}$

IV. PRACTICAL TECHNIQUES

Objective: Measure I_{DSS} and $V_{GS(off)}$ of an N-channel JFET.

Steps	Comments
1. Construct the circuit that is shown in Fig. 14-10. 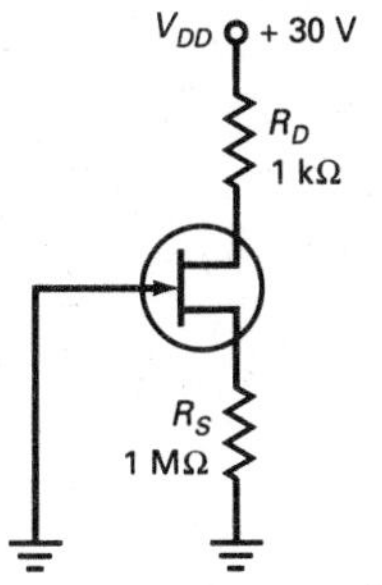 V_{DD} ○ + 30 V R_D 1 kΩ R_S 1 MΩ **Figure 14-10**	a. V_{DD} should be set to a voltage that is less than the breakdown voltage of the JFET. b. Typical values: $V_{DD} = 30$ V $R_D = 1$ kΩ $R_S = 1$ MΩ

2. Determine $V_{GS(off)}$. 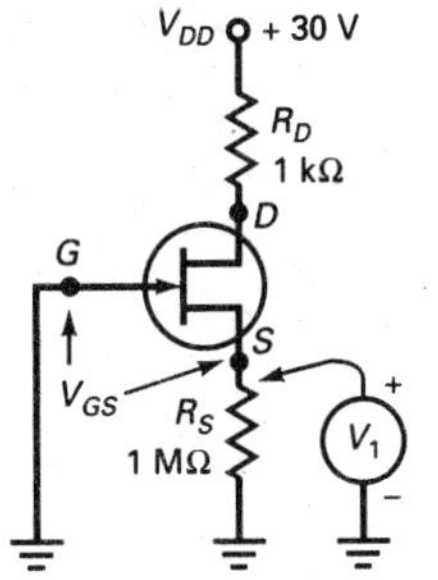
 	
V_{DD} ○ + 30 V
R_D 1 kΩ
R_S 1 MΩ, V_1
 	
Figure 14-11
 	
$V_{GS(off)} \approx -V_1$
	a. A very large value of R_S will cause the self-bias line to intercept the transconductance curve at a value close to $V_{GS(off)}$.
	b. Record the reading of V_1.
	c. $V_G = 0$ $V_S = V_1$
	   $V_{GS} = V_G - V_S = 0 - V_1 = -V_1$
	d. $I_D = \dfrac{V_1}{1 \text{ M}\Omega} \approx 0$
	   Since $I_D \approx 0$, $V_{GS} \approx V_{GS(off)}$
	e. Determine $V_{GS(off)}$ using the voltmeter reading.

3. Determine I_{DSS}. 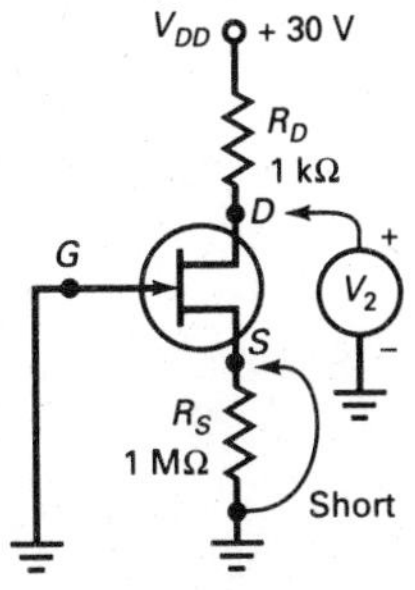
 	
V_{DD} ○ + 30 V
R_D 1 kΩ, V_2
R_S 1 MΩ, Short
 	
Figure 14-12
 	
$I_{DSS} = \dfrac{30 \text{ V} - V_2}{1 \text{ k}\Omega}$
	a. Short the source to ground by connecting a jumper across R_S.
	   $V_G = 0$ $V_S = 0$ $V_{GS} = 0$
	b. When measuring I_{DSS} be sure that $V_{DS} > V_P$.
	   $V_P = -V_{GS(off)}$ $V_{DS} = V_2$
	c. V_{DS} can be kept greater than V_P by reducing the value of R_D.
	d. Record the reading of V_2.
	e. $I_D = I_{DSS}$, since $V_{GS} = 0$ and $V_{DS} > V_P$.
	f. Determine I_{DSS} using Ohm's Law.

Example 1. From the voltmeter readings, determine the values of I_{DSS} and $V_{GS(off)}$ of the *n*-channel JFET.

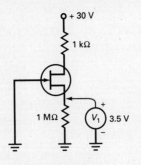

Figure 14-13

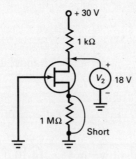

Figure 14-14

$V_{GS(off)} = -3.5 \text{ V}$

$I_{DSS} = \dfrac{30 \text{ V} - 18 \text{ V}}{1 \text{ k}\Omega} = 12 \text{ mA}$

V. PROBLEMS

Sec. 14-1 Self-Bias of JFETs

14-1. From the dc voltmeter reading for the circuit of Fig. 14-15, find the following:
  a. I_D
  b. V_G
  c. V_{GS}
  d. V_{DS}

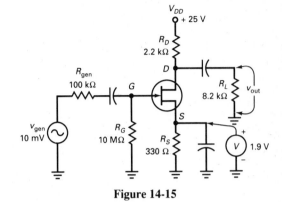

Figure 14-15

14-2. If the dc voltage between the source and gate is 2 V in the circuit of Fig. 14-16, what are the values of the following:
  a. I_D
  b. V_{DS}
  c. P_D

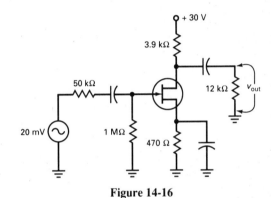

Figure 14-16

Sec. 14-2 Graphical Solution for Self-Bias

14-3. For the JFET in Fig. 14-17a and its transconductance curve shown in Fig. 14-17b, find the following:
  a. I_D b. V_{GS} c. V'_P d. V_{DS}

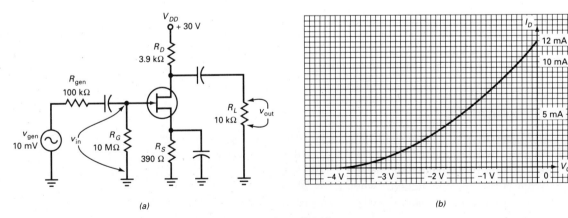

Figure 14-17

14-4. What value of R_S would cause I_D to equal 3 mA in the circuit of Fig. 14-17a?

14-5. For the circuit of Fig. 14-18 and the JFET transconductance curve of Fig. 14-17b, do the following:
  a. Complete Table 14-2 for the given circuit conditions.

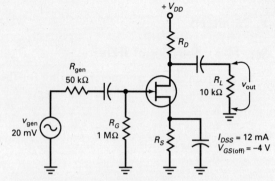

Figure 14-18

R_S Ω	V_{DD} V	R_D kΩ	I_D mA	V_{DS} V
330	24	2.7		
470	24	2.7		
330	30	2.7		
330	24	3.9		

Table 14-2

Select either *increased*, *decreased*, or *stayed the same* as the correct answer to the following questions.
  b. What happened to I_D when only R_S was increased?
  c. What happened to V_{DS} when only R_S was increased?
  d. What happened to I_D when only V_{DD} was increased?
  e. What happened to V_{DS} when only V_{DD} was increased?
  f. What happened to I_D when only R_D was increased?
  g. What happened to V_{DS} when only R_D was increased?

14-6. For the circuit of Fig. 14-19, find the following:
  a. I_D
  b. V_{GS}
  c. V'_P
  d. V_{DS}

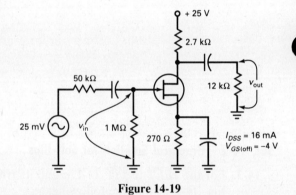

Figure 14-19

Sec. 14-3 Solution with the Universal JFET Curve

14-7. If $I_{DSS} = 10$ mA and $V_{GS(off)} = -4$ V for the JFET in the circuit of Fig. 14-17a, what are the values of the following:
  a. I_D b. V_{GS} c. V'_P d. V_{DS}

14-8. If $I_{DSS} = 15$ mA and $V_{GS(off)} = -3$ V for the JFET in the circuit of Fig. 14-19, what are the values of the following:
  a. I_D b. V_{GS} c. V'_P d. V_{DS}

Sec. 14-4 Transconductance

14-9. A JFET has $I_{DSS} = 20$ mA and $V_{GS(off)} = -4$ V.
  a. What is its transconductance when $V_{GS} = 0$ V?
  b. What is its transconductance when $V_{GS} = -2$ V?
  c. What is its transconductance when $I_D = 20$ mA?
  d. What is its transconductance when $I_D = 10$ mA?

14-10. With respect to Prob. 14-3, find the following:
  a. g_{m0}
  b. g_m

Name _____ Date _____

14-11. A JFET with $I_{DSS} = 10$ mA and $V_{GS(off)} = -5$ V is operating with $g_m = 2000$ μmho.
    a. What is the value of V_{GS}?
    b. What is the value of I_D?

Sec. 14-5   JFET Amplifiers

14-12. The circuit in Fig. 14-20 is an ac equivalent circuit of a common source amplifier. If $g_m = 4000$ μmho, find the following:
    a. Z_{in}
    b. v_{in}
    c. A
    d. v_{out}
    e. i_d
    f. v_{gs}

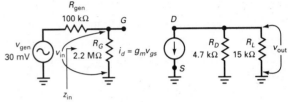

Figure 14-20

14-13. For the ac equivalent circuit of a CS amplifier in Fig. 14-21, do the following:
    a. Complete Table 14-3 for the given circuit conditions.

g_m μmho	R_L kΩ	A	v_{in} mV
3000	5.6		
4000	5.6		
3000	12		

Table 14-3

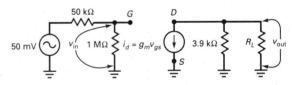

Figure 14-21

    Select either *increased*, *decreased*, or *stayed the same* as the correct answer to the following questions.
    b. What happened to the gain when only g_m was increased?
    c. What happened to the gain when only R_L was increased?
    d. What happened to v_{in} when only g_m was increased?
    e. What happened to v_{in} when only R_L was increased?

14-14. For the circuit of Fig. 14-17a and the JFET transconductance curve of Fig. 14-17b, find the following:
    a. Z_{in} **b.** v_{in} **c.** A **d.** v_{out} **e.** i_d **f.** v_{gs}

14-15. What is the ac output voltage for the circuit of Fig. 14-19?

14-16. The circuit in Fig. 14-22 is an ac equivalent circuit of a source follower. If $g_m = 3000$ μmho, find the following:
    a. Z_{in}
    b. v_{in}
    c. A
    d. v_{out}
    e. i_d
    f. v_{gs}

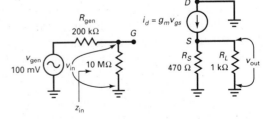

Figure 14-22

14-17. From the dc voltmeter reading in the circuit of Fig. 14-23, find the following:
a. g_m
b. v_{in}
c. A
d. v_{out}
e. i_d
f. v_{gs}

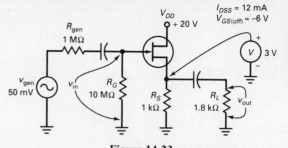

Figure 14-23

14-18. The JFET of the SF of Fig. 14-24b has the transconductance curve of Fig. 14-24a. What is the ac output voltage?

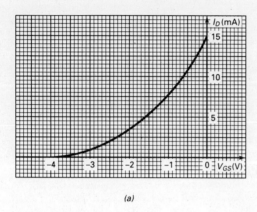

(a)   (b)

Figure 14-24

Sec. 14-6 The JFET Analog Switch

14-19. The input voltage of the circuit in Fig. 14-25 is 100 mV peak to peak.
a. What is the output voltage when $V_G = 0$ V?
b. What is the output voltage when $V_G = -5$ V?

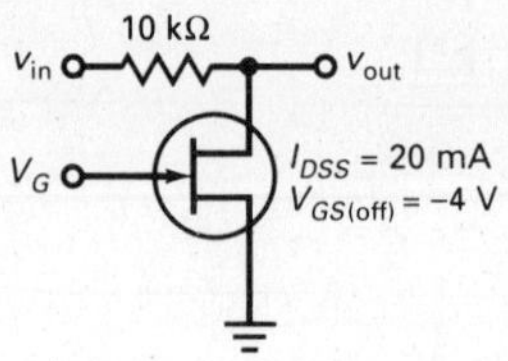

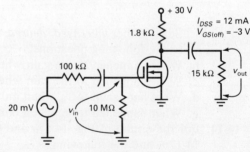

Figure 14-25

Sec. 14-7 Depletion-Mode MOSFET Amplifier

14-20. For the circuit of Fig. 14-26, find the following:
a. v_{in}
b. A
c. v_{out}

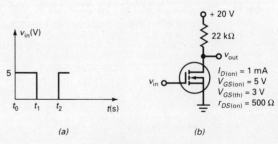

Figure 14-26

Sec. 14-8 Enhancement-Mode MOSFET Application

14-21. For the circuit of Fig. 14-27, determine the following:
a. What is v_{out} between the time intervals t_0 and t_1?
b. What is v_{out} between the time intervals t_1 and t_2?

(a)   (b)

Figure 14-27

Sec. 14-9 Other JFET Biasing

14-22. For the circuit of Fig. 14-28, find the following:
- a. I_D
- b. V_G
- c. V_{GS}
- d. V_S
- e. V_D
- f. V_{DS}
- g. g_{m0}
- h. g_m

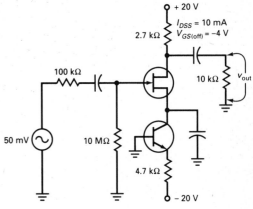

Figure 14-28

CHAPTER 15
THYRISTORS

Study Chap. 15 in *Electronic Principles*.

I. TRUE / FALSE

Answer true (T) or false (F) to each of the following statements.

Answer

1. A four-layer diode is basically made up of two diodes. (15-1) 1. ___
2. Because the switching action of a thyristor is based on positive feedback, it is called a latch. (15-1) 2. ___
3. The simplest type of thyristor can only be closed by breakover and opened by low-current dropout. (15-1) 3. ___
4. The most popular types of thyristors are closed by triggering and are opened by low-current dropout. (15-1) 4. ___
5. The only way to close a four-layer diode that has three external terminals is by breakover. (15-1) 5. ___
6. The ideal approximation of a four-layer diode is an opened switch when nonconducting and a closed switch when conducting. (15-1) 6. ___
7. Most silicon controlled rectifiers are designed for trigger closing and trigger opening. (15-2) 7. ___
8. Almost all SCRs are low-current devices. (15-2) 8. ___
9. If the SCR's critical rate of voltage rise is exceeded, it can be falsely triggered. (15-2) 9. ___
10. If the SCR's critical rate of current rise is exceeded, it may be destroyed. (15-2) 10. ___
11. After the SCR has been turned on, a way to reset it is by reducing its anode current to a value that is less than its holding current. (15-2) 11. ___
12. The gate of a photo-SCR should be left opened for it to have a maximum sensitivity to light. (15-3) 12. ___
13. After a light trigger has closed the photo-SCR, it remains closed, even though the light disappears. (15-3) 13. ___
14. The silicon controlled switch is a high-power device like an SCR. (15-3) 14. ___
15. Latch current in a diac can only flow in one direction. (15-4) 15. ___
16. Triacs are used to control the average load current in industrial heating, lighting, and other heavy-power applications. (15-4) 16. ___
17. The diac is nonconducting until the voltage across it tries to exceed the breakover voltage in either direction. (15-4) 17. ___

18. The number of doped regions in a unijunction transistor is three. (15-5)   18. ___
19. In order for emitter current to flow in a UJT, the emitter voltage must be greater than intrinsic standoff voltage. (15-5)   19. ___
20. Troubleshooting at a system level involves isolating faults using block diagrams. (15-6)   20. ___

II. COMPLETION

Complete each of the following statements.   *Answer*

1. A sharp pulse that causes the transistor latch to close is called a ___. (15-1)   1. _____
2. The method of using a large enough supply voltage to cause a transistor latch to close is called ___. (15-1)   2. _____
3. A four-layer diode that has only two external leads is also known as a ___ diode. (15-1)   3. _____
4. The internal transistors of a four-layer diode go into ___ when the latch is closed. (15-1)   4. _____
5. The latch will open in a four-layer diode when the anode current is reduced below the ___ current. (15-1)   5. _____
6. The second approximation sets the knee voltage of a four-layer diode to ___ V. (15-1)   6. _____
7. The number of external terminals that an SCR has is ___. (15-2)   7. _____
8. On data sheets, the SCR's breakover voltage is often called the forward ___ voltage. (15-2)   8. _____
9. One way to reduce the effects of switching transients that may cause an excessive rise in the voltage rate for an SCR is by using a ___ snubber. (15-2)   9. _____
10. One way to reduce an excessive rise in the current rate for an SCR is to put an ___ in series with it. (15-2)   10. _____
11. A circuit that uses an SCR to short the load terminals at the first sign of overvoltage is called a ___ . (15-2)   11. _____
12. A photo-SCR is also known as a ___-___ SCR. (15-3)   12. _____
13. An SCR that is closed by a positive trigger and opened by a negative trigger is called a ___-controlled switch. (15-3)   13. _____
14. The number of triggering terminals that an SCS has is ___. (15-3)   14. _____
15. A thyristor that acts like two SCRs in parallel is called a ___. (15-4)   15. _____
16. The number of external terminals on a diac is ___. (15-4)   16. _____
17. Triacs are normally closed by a ___ voltage. (15-4)   17. _____
18. The quantity η that is defined for a UJT is called a ___ standoff ratio. (15-5)   18. _____
19. The symbol I_v for a UJT that is equivalent to a holding current is called a ___ current. (15-5)   19. _____
20. Things that are being done by different parts of the overall circuit can be represented by ___ blocks. (15-6)   20. _____

III. ILLUSTRATIVE AND PRACTICE PROBLEMS

Note: Illustrative Problems 1 through 3 are only an exercise in determining the switching characteristics of a Shockley diode, SCR, and UJT. The effects of any critical rates have been neglected.

Illustrative Problem 1. The dc voltage source V_{CC} is varied (see Fig. 15-1a). Find v_{out} and the current I through the Shockley diode during the following time intervals:

a. $t_0 < t < t_1$
b. $t_1 < t < t_2$
c. $t_2 < t < t_3$

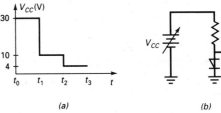

(a)   (b)

Figure 15-1

Steps	Comments
1. Find v_{out} and I. ($t_0 < t < t_1$) $I = \dfrac{30\text{ V} - 0.7\text{ V}}{1\text{ k}\Omega} = 29.3\text{ mA}$ Since $I > I_H$ ($I_H = 5$ mA) $I = 29.3$ mA $v_{out} = 0.7$ V	a. When the latch is opened, $v_{out} = V_{CC}$, $I = 0$ mA. b. When the latch is closed, $v_{out} = 0.7$ V. c. The switching current is assumed to be zero. d. Since the latch was opened prior to this time interval, it will remain open until the following conditions are met: $$V_{CC} \geq V_B \quad \text{and} \quad I \geq I_H$$ e. $V_{CC} = 30$ V. Since $V_{CC} > V_B$, assume that the latch closes and solve for I. f. In order for your assumption to be correct, $I \geq I_H$. g. Calculations indicate the latch closes.
2. Find v_{out} and I. ($t_1 < t < t_2$) $I = \dfrac{10\text{ V} - 0.7\text{ V}}{1\text{ k}\Omega} = 9.3\text{ mA}$ Since $I > I_H$ ($I_H = 5$ mA) $I = 9.3$ mA $v_{out} = 0.7$ V	a. $V_{CC} = 10$ V. b. Since the latch was closed prior to this time interval, it will remain closed until I becomes less than I_H. c. Assume the latch is closed and solve for I. d. Calculations indicate the latch stays closed.
3. Find v_{out} and I. ($t_2 < t < t_3$) $I = \dfrac{4\text{ V} - 0.7\text{ V}}{1\text{ k}\Omega} = 3.3\text{ mA}$ Since $I < I_H$ ($I_H = 5$ mA) $I = 0$ mA $v_{out} = 4$ V open circuit voltage	a. $V_{CC} = 4$ V. b. Since the latch was closed prior to this time interval, it will remain closed until I becomes less than I_H. c. Assume the latch is closed and solve for I. d. Calculations indicate the latch opens. e. Determine v_{out} and I for an opened latch.

Copyright © 1993 by the Glencoe Division of Macmillan/McGraw-Hill School Publishing Company. All rights reserved.

Practice Problem 1. The dc voltage source V_{CC} is varied (see Fig. 15-2a). Find v_{out} and the current I through the Shockley diode during the following time intervals:

a. $t_0 < t < t_1$
b. $t_1 < t < t_2$
c. $t_2 < t < t_3$

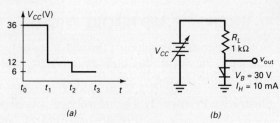

(a)　　　　(b)

Figure 15-2

Answers: a. $v_{out} = 0.7$ V, $I = 35.3$ mA　　b. $v_{out} = 0.7$ V, $I = 11.3$ mA　　c. $v_{out} = 6$ V, $I = 0$ mA

Illustrative Problem 2. The dc voltage sources V_{CC} and V_{in} are varied (see Fig. 15-3a). Find v_{out} and the current I through the SCR diode during the following time intervals:

a. $t_0 < t < t_1$
b. $t_1 < t < t_2$
c. $t_2 < t < t_3$
d. $t_3 < t < t_4$

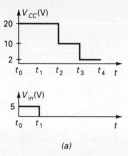
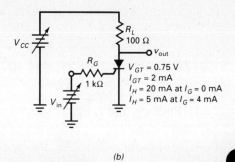

(a)　　　　(b)

Figure 15-3

Steps	Comments
1. Find v_{out} and I. ($t_0 < t < t_1$) $I_G = \dfrac{5\text{V} - 0.75\text{V}}{1\text{ k}\Omega} = 4.25$ mA $I_G > I_{GT}$　($I_{GT} = 2$ mA) $I = \dfrac{20\text{ V} - 0.7\text{ V}}{100\ \Omega} = 193$ mA Since $I > I_H$　($I_H = 5$ mA at $I_G = 4$ mA) $I = 193$ mA　　$v_{out} = 0.7$ V	a. When the latch is opened, $v_{out} = V_{CC}$, $I = 0$ mA. b. When the latch is closed, $v_{out} = 0.7$ V. c. The holding current is reduced when the gate current is increased. d. Since the latch was opened prior to this time interval, it will remain open until the following conditions are met: $I_G \geq I_{GT}$　and　$I \geq I_H$ e. Solve for I_G. f. If $I_G \geq I_{GT}$, assume that the latch closes and solve for I. g. In order for your assumption to be correct, $I \geq I_H$. h. Calculations indicate the latch closes.
2. Find v_{out} and I. ($t_1 < t < t_2$) $I = \dfrac{20\text{ V} - 0.7\text{ V}}{100\ \Omega} = 193$ mA Since $I > I_H$　($I_H = 20$ mA at $I_G = 0$ mA) $I = 193$ mA　　$v_{out} = 0.7$ V	a. Since $v_{in} = 0$ V, $I_G = 0$ mA. b. As long as $I \geq I_H$, the latch will remain closed. c. Since there is no change in the value of V_{CC}, the latch remains closed with the same values of v_{out} and I as in step 1.

148　Copyright © 1993 by the Glencoe Division of Macmillan/McGraw-Hill School Publishing Company. All rights reserved.

Steps	Comments
3. Find v_{out} and I. ($t_2 < t < t_3$) $I = \dfrac{10\text{ V} - 0.7\text{ V}}{100\ \Omega} = 93\text{ mA}$ Since $I > I_H$ ($I_H = 20$ mA at $I_G = 0$ mA) $I = 93$ mA $v_{out} = 0.7$ V	a. Since $v_{in} = 0$ V, $I_G = 0$ mA. b. As long as $I \geq I_H$, the latch will remain closed. c. Since the value of V_{CC} changes, assume that the latch is still closed and solve for I. d. Calculations indicate the latch stays closed.
4. Find v_{out} and I. ($t_3 < t < t_4$) $I = \dfrac{2\text{ V} - 0.7\text{ V}}{100\ \Omega} = 13\text{ mA}$ Since $I < I_H$ ($I_H = 20$ mA at $I_G = 0$ mA) $I = 0$ mA $v_{out} = 2$ V open circuit voltage	a. Since $v_{in} = 0$ V, $I_G = 0$ mA. b. As long as $I \geq I_H$, the latch will remain closed. c. Since the value of V_{CC} changes, assume that the latch is still closed and solve for I. d. Calculations indicate the latch opens. e. Determine v_{out} and I for an opened latch.

Practice Problem 2. The dc voltage sources V_{CC} and V_{in} are varied (see Fig. 15-4a). Find v_{out} and the current I through the SCR diode during the following time intervals:

a. $t_0 < t < t_1$
b. $t_1 < t < t_2$
c. $t_2 < t < t_3$
d. $t_3 < t < t_4$

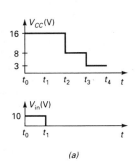
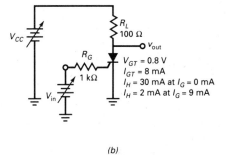

Figure 15-4

Answers:
a. $v_{out} = 0.7$ V, $I = 153$ mA   c. $v_{out} = 0.7$ V, $I = 73$ mA
b. $v_{out} = 0.7$ V, $I = 153$ mA   d. $v_{out} = 3$ V, $I = 0$ mA

Illustrative Problem 3. The V_{EE} voltage source is varied (see Fig. 15-5a). Using the second approximation, find V_E and the emitter current I_E through the UJT during the following time intervals (assume that $V_{E(sat)}$ is constant):

a. $t_0 < t < t_1$
b. $t_1 < t < t_2$
c. $t_2 < t < t_3$

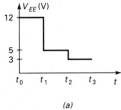
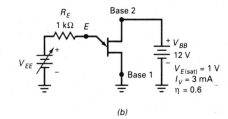

Figure 15-5

Steps	Comments
1. Find V_E and I_E. ($t_0 < t < t_1$) *(Figure 15-6)* $\eta V_{BB} + 0.7\text{ V} = (0.6)(15\text{ V}) + 0.7\text{ V} = 9.7\text{ V}$ $V_{EE} = 12\text{ V} \quad V_{EE} > 9.7\text{ V}$ $I_E = \dfrac{12\text{ V} - 1\text{ V}}{1\text{ k}\Omega} = 11\text{ mA}$ Since $I_E > I_V \quad (I_V = 3\text{ mA})$ $V_E = V_{E(\text{sat})} = 1\text{ V}$	a. Fig. 15-6 is an equivalent circuit. b. When the latch is opened, $V_E = V_{EE}$, $I_E = 0$ mA. c. When the latch is closed, $V_E = V_{E(\text{sat})}$. d. Since the latch was opened prior to this time interval, it will remain open until the following conditions are met: $\quad V_{EE} > \eta V_{BB} + 0.7\text{ V} \quad$ and $\quad I_E \geq I_V$ e. Solve for the value of $\eta V_{BB} + 0.7$ V. f. If $V_{EE} > \eta V_{BB} + 0.7$ V, assume that the latch closes and solve for I_E. $V_E = V_{E(\text{sat})} = 1$ V g. In order for your assumption to be correct, $I_E > I_V$. h. Calculations indicate the latch closes and R_1 becomes very small.
2. Find V_E and I_E. ($t_1 < t < t_2$) $I_E = \dfrac{5\text{ V} - 1\text{ V}}{1\text{ k}\Omega} = 4\text{ mA}$ Since $I_E > I_V \quad (I_V = 3\text{ mA})$ $V_E = V_{E(\text{sat})} = 1\text{ V}$	a. As long as $I_E > I_V$, the latch will remain closed. b. Since the value of V_{EE} changes, assume that the latch is still closed and solve for I_E. c. Calculations indicate the latch stays closed.
3. Find V_E and I_E. ($t_2 < t < t_3$) $I_E = \dfrac{3\text{ V} - 1\text{ V}}{1\text{ k}\Omega} = 2\text{ mA}$ Since $I_E < I_V \quad (I_V = 3\text{ mA})$ $I_E = 0\text{ mA} \quad V_E = V_{EE} = 3\text{ V}$ open circuit voltage	a. As long as $I_E > I_V$, the latch will remain closed. b. Since the value of V_{EE} changes, assume that the latch is still closed and solve for I_E. c. Calculations indicate the latch opens. d. Determine V_E and I_E for an opened latch.

Practice Problem 3. The V_{EE} voltage source is varied (see Fig. 15-7a). Using the second approximation, find V_E and the emitter current I_E through the UJT during the following time intervals (assume that $V_{E(\text{sat})}$ is constant):

a. $t_0 < t < t_1$
b. $t_1 < t < t_2$
c. $t_2 < t < t_3$

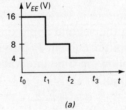
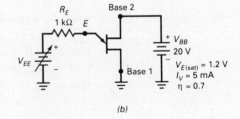

(a)   (b)

Figure 15-7

Answers: a. $V_E = 1.2$ V, $I_E = 14.8$ mA   b. $V_E = 1.2$ V, $I_E = 6.8$ mA   c. $V_E = 4$ V, $I_E = 0$ mA

Name ______________________________ Date ______________

IV. PROBLEMS

Note: Unless otherwise indicated, use the second approximation method to determine the answers to all the problems in this chapter.

Sec. 15-1 The Four-Layer Diode

15-1. For the circuit of Fig. 15-8, determine the following:

- **a.** The minimum battery voltage that would turn the latch on.
- **b.** The current through the diode when the battery voltage is 20 V.
- **c.** The minimum battery voltage that would keep the latch on once the Shockley diode is conducting.

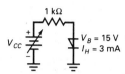

Figure 15-8

15-2. For the circuit of Fig. 15-9, determine the following:

- **a.** V_A and the current through the diode when $V_{CC} = 30$ V.
- **b.** V_A and the current through the diode when V_{CC} is reduced from 30 V to 8 V.
- **c.** V_A and the current through the diode when V_{CC} is reduced from 30 V to 4 V.

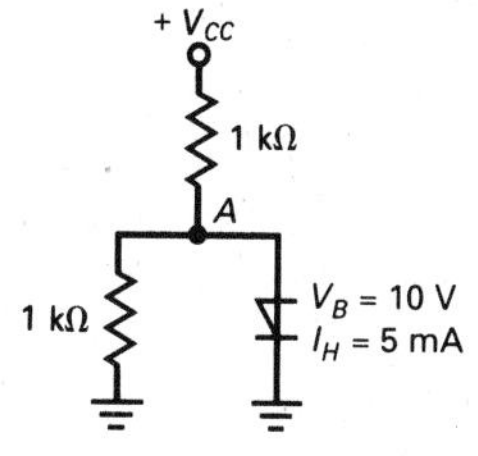

Figure 15-9

15-3. For the circuit and the output voltage waveform of Fig. 15-10, determine the following:

- **a.** V_{max}
- **b.** V_{min}
- **c.** Current through the 10-kΩ resistor at $t = t_0$
- **d.** Current through the 10-kΩ resistor at $t = t_1$
- **e.** Current through the 10-kΩ resistor at $t = t_2$
- **f.** Current through the Shockley diode at $t = t_2$

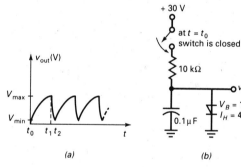

Figure 15-10

15-4. For the circuit and the output voltage waveform of Fig. 15-11, determine the following:

- **a.** V_{max}
- **b.** V_{min}
- **c.** Current through the 1-kΩ resistor at $t = t_0$
- **d.** Current through the 1-kΩ resistor at $t = t_1$
- **e.** Current through the 1-kΩ resistor at $t = t_2$
- **f.** Current through the Shockley diode at $t = t_2$

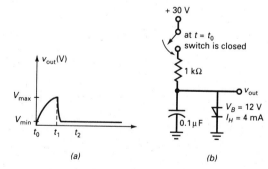

Figure 15-11

15-5. For the circuit and the output voltage waveform of Fig. 15-12, determine the following:
  a. V_1 b. V_2 c. V_3

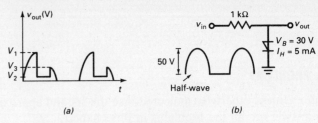

Figure 15-12

Sec. 15-2 The Silicon Controlled Rectifier

15-6. The SCR of Fig. 15-13 has $V_{GT} = 0.74$ V, $I_{GT} = 3$ mA, $I_H = 2$ mA, and $I_H = 4$ mA at $I_G = 0$ mA. Determine the following:
  a. The output voltage when the SCR is off.
  b. The output voltage when the latch is closed.
  c. The minimum input voltage that will trigger the SCR.

15-7. The SCR of Fig. 15-14 has $V_{GT} = 1$ V, $I_{GT} = 4$ mA, $I_H = 2$ mA, and $I_H = 5$ mA at $I_G = 0$ mA. At $t = t_0$ the switch is closed. At $t = t_1$ (sometime later) the switch is re-opened. Find the following:
  a. I_G, I_L, and v_{out} (before the switch is closed)
  b. I_G, I_L, and v_{out} (after the switch is closed)
  c. I_G, I_L, and v_{out} (after the switch is re-opened)

15-8. Repeat Prob. 15-7 with $R_G = 4.7$ kΩ.

15-9. Repeat Prob. 15-7 with $R_L = 3.3$ kΩ.

15-10. The SCR of Fig. 15-15 has $V_{GT} = 0.76$ V, $I_{GT} = 3$ mA, $I_H = 2$ mA, and $I_H = 5$ mA at $I_G = 0$ mA. Determine the following:
  a. The value of R_G that would limit the gate current to 3 mA.
  b. The value of R_L that would limit the current through the SCR to 10 mA.

15-11. The zener diode in Fig. 15-16 has a breakdown voltage of 10 V. Calculate the supply voltage that turns on the crowbar. (Neglect the zener resistance.)

15-12. The switch in the circuit of Fig. 15-17 is closed at $t = t_0$. Determine the following:
  a. The voltage across the capacitor an instant before the SCR fires.
  b. The voltage across the capacitor after it discharges.
  c. The current through the 1-kΩ resistor when the SCR fires.
  d. The current through the 1-kΩ resistor after the capacitor discharges.
  e. The current through the 1.5-kΩ resistor after the SCR fires.

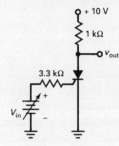

Figure 15-13

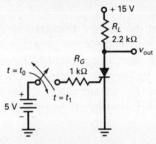

Figure 15-14

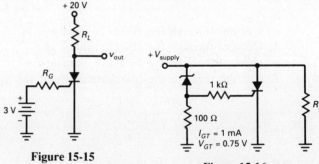

Figure 15-15   Figure 15-16

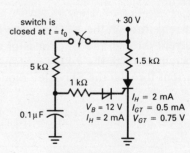

Figure 15-17

Sec. 15-3 Variations of the SCR

15-13. The circuit of Fig. 15-18 is in a dark room. When a bright light is turned on, the LASCR was closed. Determine the following:
   a. The current through the LASCR before the light is turned on.
   b. The current through the LASCR after the light is turned on.
   c. The current through the LASCR if the light is turned off.
   d. Would the value of R_G have to be increased or decreased to cause the LASCR sensitivity to light to increase?

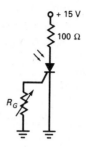

Figure 15-18

Sec. 15-4 Bidirectional Thyristors

15-14. For the circuit and waveforms of Fig. 15-19, determine the following:
   a. I_1 **b.** I_2 **c.** I_3

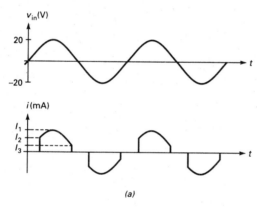
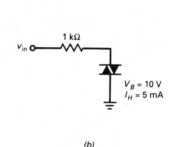

(a) (b)

Figure 15-19

15-15. Sketch the current waveform for the circuit and voltage waveforms of Fig. 15-20. (Assume the gate pulse will cause the triac to fire and the holding current is zero.)

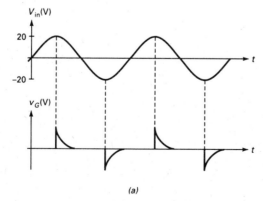
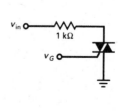

(a) (b)

Figure 15-20

Sec. 15-5 The Unijunction Transistor

15-16. From the equivalent circuit of the UJT in Fig. 15-21, determine the following:
  a. The intrinsic standoff ratio.
  b. The intrinsic standoff voltage.
  c. The minimum voltage that would have to be applied to the emitter to cause the emitter diode to turn on. (Allow 0.7 V across the emitter diode.)
  d. What happens to the value of R_1 when the emitter diode is turned on?

15-17. For the UJT of Fig. 15-22 that has an η of 0.65, do the following:
  a. Find the minimum value of V that turns on the UJT. (Allow 0.7 V across the emitter diode.)
  b. Find the value of V that will just turn off the UJT after it has been turned on.

15-18. For the circuit and output voltage waveform of Fig. 15-23, determine the following (allow 0.7 V across the emitter diode):
  a. V_1 b. V_2

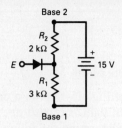

Figure 15-21

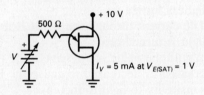

Figure 15-22

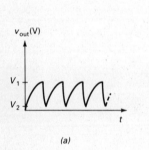 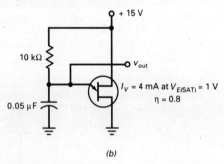

Figure 15-23

Name _____ Date _____

CHAPTER 16
FREQUENCY EFFECTS

Study Chap. 16 in *Electronic Principles*.

I. TRUE / FALSE

Answer true (T) or false (F) to each of the following statements.

Answer

1. The graph of the frequency response of an amplifier shows how the output voltage of the amplifier is affected by changes in frequency. **(16-1)**   1. ___
2. At high frequencies, the output voltage of an amplifier decreases because of transistor and stray-wiring capacitance. **(16-1)**   2. ___
3. As the frequency of a generator increases, the output voltage of a coupling circuit approaches its minimum value. **(16-2)**   3. ___
4. The resistance value needed to calculate the critical frequency of the input coupling circuit of a CE amplifier is equal to the sum of the generator's resistance and the amplifier's input impedance. **(16-2)**   4. ___
5. The resistance value needed to calculate the critical frequency of the output coupling circuit of a CE amplifier is equal to the parallel combination of the load resistance and the amplifier's output impedance. **(16-3)**   5. ___
6. The frequency response of an output coupling circuit is similar to an input coupling circuit. **(16-3)**   6. ___
7. When the generator frequency is equal to the critical frequency produced by the emitter bypass capacitor, the emitter is at ac ground. **(16-4)**   7. ___
8. The lowest frequency in the midband of a CE amplifier is 10 times greater than the highest of the emitter bypass and coupling circuit's critical frequencies. **(16-4)**   8. ___
9. The internal capacitance of a transistor is typically in microfarads. **(16-5)**   9. ___
10. At high frequencies, the output side of the transistor acts as an ordinary bypass circuit with an *R* and a *C*. **(16-5)**   10. ___
11. The Miller theorem permits you to replace a feedback capacitor that is between the input and output of an inverting amplifier with two equivalent capacitors, one on the input side and one on the output side. **(16-6)**   11. ___
12. The output Miller equivalent capacitor of a CE amplifier is much greater than the feedback capacitor of the transistor. **(16-7)**   12. ___
13. When a CE amplifier is operating below its midband, the coupling capacitors and emitter bypass capacitors cause the output voltage to decrease. **(16-7)**   13. ___
14. The base spreading resistor r_b' has a large effect on the frequency response below the midband. **(16-7)**   14. ___

15. The output voltage outside of the midband will always be less than its maximum value. **(16-8)**
16. Each time the ordinary power gain increases by a factor of 10, the decibel power gain increases by 3 dB. **(16-9)**
17. Each time the ordinary voltage gain increases by a factor of 2, the decibel voltage gain increases by 6 dB. **(16-10)**
18. A decade below 1000 Hz is 900 Hz. **(16-11)**
19. A rate of 20 dB per decade is equal to 6 dB per octave. **(16-11)**

15. ___
16. ___
17. ___
18. ___
19. ___

II. COMPLETION

Complete each of the following statements.

Answer

1. The band of frequencies where the amplifier is producing the maximum output voltage is called ____. **(16-1)**
2. When an amplifier is operating at its critical frequencies, the value of its output voltage is ____ percent of its maximum value. **(16-1)**
3. The amplifier is producing maximum output voltage when the frequency it is operating at is between 10 times its lower critical frequency and ____ times its upper critical frequency. **(16-1)**
4. The coupling and the emitter bypass capacitors are responsible for the amplifier's ____ critical frequency. **(16-1)**
5. When the generator frequency is 10 times greater than a coupling circuit's critical frequency, the coupling circuit's current is within ____ percent of its maximum value. **(16-2)**
6. If the coupling capacitance is increased, the critical frequency of the coupling circuit will ____. **(16-2)**
7. If the value of the load resistance is increased, the critical frequency of the output coupling circuit will ____. **(16-3)**
8. In order to find the critical frequency produced by the emitter bypass capacitor, the output impedance of the ____ must first be determined. **(16-4)**
9. When determining the critical frequency produced by the emitter bypass capacitor of a CE amplifier, you can assume that the input and output coupling capacitors are ____ shorts. **(16-4)**
10. The internal capacitance of a transistor should be taken into account when the amplifier is operating above ____ kHz. **(16-5)**
11. The unwanted capacitance that exists between the leads and chassis of a circuit is called ____ wiring capacitance. **(16-5)**
12. A capacitor that is connected between the input and output terminals of an amplifier is called a ____ capacitor. **(16-6)**
13. The voltage gain used to calculate the Miller equivalent capacitance is the ____ voltage gain of the amplifier. **(16-6)**
14. Of the two critical frequencies that are above the midband in an amplifier, the one closer to the midband is more important and is called the ____ critical frequency. **(16-7)**
15. The frequency at which the current gain β of the transistor drops to unity is called the current gain-bandwidth ____. **(16-7)**
16. Each time the ordinary power gain increases by a factor of 2, the decibel power gain increases by ____ dB. **(16-9)**
17. Each time the ordinary voltage gain increases by a factor of 10, the decibel voltage gain increases by ____ dB. **(16-10)**
18. The graph of the voltage gain in decibels versus frequency plotted on semilogarithmic paper is called a ____ plot. **(16-11)**
19. An octave above 200 Hz is ____. **(16-11)**

III. ILLUSTRATIVE AND PRACTICE PROBLEMS

Illustrative Problem 1. Find the critical frequency of the input coupling circuit.

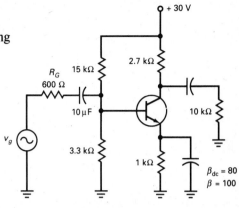

Figure 16-1

Steps	Comments
1. Draw the input coupling circuit. Figure 16-2	In order to find the critical frequency of the input coupling circuit, the Thevenin resistance with respect to the capacitor's terminals must be known. However, it is necessary to find the value of z_{in} before the value of r_{th} can be found.
2. Find z_{in}. $z_{in} = 444\ \Omega$	An earlier analysis of this circuit was done in Chaps. 9 and 10 and yielded the value of z_{in}.
3. Find r_{th}. Figure 16-3 $r_{th} = 600\ \Omega + 444\ \Omega = 1044\ \Omega$	a. Remove the capacitor. b. Replace the voltage source with a short. c. Calculate the Thevenin resistance between the terminals. $$r_{th} = R_G + z_{in}$$
4. Find f_c. $$f_c = \frac{1}{2\pi(1044\ \Omega)10\ \mu F} = 15.2\ Hz$$	The value of R is equal to the Thevenin resistance r_{th}. $$f_c = \frac{1}{2\pi RC}$$

Practice Problem 1. Find the critical frequency of the input coupling circuit. (See Prac. Prob. 9-2.)

Answer: $f_c = 126$ Hz

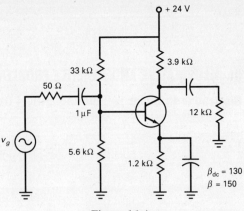

Figure 16-4

Illustrative Problem 2. What is the input and output Miller capacitance of the amplifier due to an external feedback capacitor?

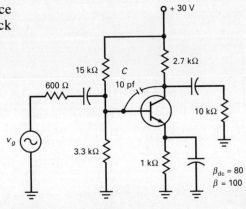

Figure 16-5

Steps	Comments
1. Redraw the circuit to show the Miller capacitance. ![Figure 16-6]	Using Miller's theorem, split the external feedback capacitor C into two equivalent capacitors, one at the input side and one at the output side of the amplifier.
2. Find $C_{in(Miller)}$. $C_{in(Miller)} = 10 \text{ pF}(401 + 1) = 4020 \text{ pF}$	$C_{in(Miller)} = C(A + 1)$ An earlier analysis of this circuit was done in Chaps. 9 and 10 and yielded $A = 401$

Steps	Comments
3. Find $C_{out(Miller)}$. $$C_{out(Miller)} = 10 \text{ pF}$$	$$C_{out(Miller)} = \frac{A+1}{A} C$$ $$\frac{A+1}{A} \approx 1$$ $$C_{out(Miller)} = C$$

Practice Problem 2. What is the input and output Miller capacitance of the amplifier due to an external feedback capacitor? (See Prac. Prob. 10-1.)

Answers: $C_{in(Miller)} = 4110$ pF, $C_{out(Miller)} = 15$ pF

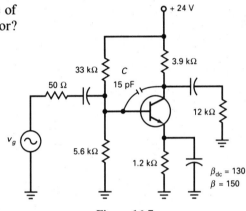

Figure 16-7

Illustrative Problem 3. Find the critical frequency of the base bypass circuit if r'_b is 106 Ω.

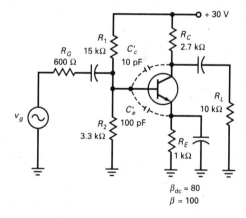

Figure 16-8

Steps	Comments
1. Find r_g. $\dfrac{1}{r_g} = \dfrac{1}{600\,\Omega} + \dfrac{1}{15\,\text{k}\Omega} + \dfrac{1}{3.3\,\text{k}\Omega}$ $r_g = 491\,\Omega$	The ac Thevenin resistance facing the base. $r_g = R_1 \,\|\, R_2 \,\|\, R_G$
2. Find the total base bypass capacitance. **Figure 16-9** $C_{in(total)} = 100\,\text{pF} + 4020\,\text{pF}$ $C_{in(total)} = 4120\,\text{pF}$	**a.** Draw the equivalent base circuit that includes r'_b, r_g, $\beta r'_e$, C'_e, and $C_{in(Miller)}$. **b.** All the capacitors that make up the base bypass capacitance are in parallel. $C_{in(total)} = C'_e + C_{in(Miller)}$ **c.** Since $C'_c = 10\,\text{pF}$, we can let $C'_c = C$ (See Illus. Prob. 16-2.) $C_{in(Miller)} = 4020\,\text{pF}$
3. Find r_{th}. $r_{th} = \dfrac{(491\,\Omega + 106\,\Omega)(531\,\Omega)}{491\,\Omega + 106\,\Omega + 531\,\Omega} = 281\,\Omega$	The ac Thevenin resistance facing the base bypass capacitance $C_{in(total)}$. $r_{th} = (r_g + r'_b) \,\|\, \beta r'_e$ $\beta r'_e = 531\,\Omega$ (See Illus. Prob. 9-2)
4. Find f_c. $f_c = \dfrac{1}{2\pi(4120\,\text{pF})(281\,\Omega)} = 137\,\text{kHz}$	$f_c = \dfrac{1}{2\pi C_{in(total)} R}$ $R = r_{th}$

Practice Problem 3. Find the critical frequency of the base bypass circuit if r'_b is 324 Ω.

Answer: $f_c = 124\,\text{kHz}$

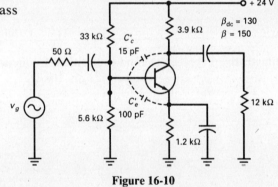

Figure 16-10

IV. PROBLEMS

Note: Assume $\beta_{dc} = \beta$ for all the problems in this chapter.

Sec. 16-1 Frequency Response of an Amplifier

16-1. An amplifier has these critical frequencies: $f_1 = 5$ Hz and $f_2 = 250$ kHz. What are the midband frequencies?

16-2. If the voltage output of an amplifier is 2 V when it is operating in the midband range, what is the voltage output when it is operating at its critical frequencies?

16-3. If an amplifier's midband range is from 20 Hz to 30 kHz, what are its critical frequencies?

Sec. 16-2 Input Coupling Capacitor

16-4. For the circuit of Fig. 16-11, do the following:
  a. Complete Table 16-1 by finding the critical frequency of the input coupling circuit for the given circuit conditions.

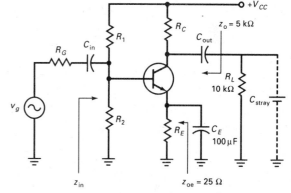

C_{in} µF	z_{in} kΩ	R_G kΩ	f_c Hz
10	4	1	
20	4	1	
10	8	1	
10	4	2	

Table 16-1

Figure 16-11

Select either *increased*, *decreased*, or *stayed the same* as the correct answer to the following questions.

  b. What happened to f_c when only C_{in} was increased?
  c. What happened to f_c when only z_{in} was increased?
  d. What happened to f_c when only R_G was increased?

16-5. If $R_G = 100$ Ω and $z_{in} = 1$ kΩ in the circuit of Fig. 16-11, what would C_{in} be for the input coupling circuit to have a critical frequency of 10 Hz?

16-6. Find the critical frequency of the input coupling circuit in Fig. 16-12 for the following circuit conditions:
  a. When $\beta = 100$.
  b. When $\beta = 200$.

16-7. Repeat prob. 16-6 for the circuit of Fig. 16-13.

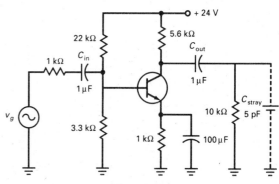

Figure 16-12

Sec. 16-3 Output Coupling Capacitor

16-8. For the circuit of Fig. 16-11, find the critical frequency of the output coupling circuit for the following values of C_{out}:
  a. 0.5 µF   b. 1 µF

16-9. In the circuit of Fig. 16-11, what would C_{out} be for the output coupling circuit to have a critical frequency of 1 Hz?

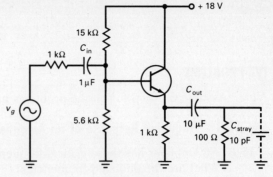

Figure 16-13

16-10. Find the critical frequency of the output coupling circuit in Fig. 16-12 when $\beta = 100$.

16-11. Find the critical frequency of the output coupling circuit in Fig. 16-13 when $\beta = 100$.

Sec. 16-4 Emitter Bypass Capacitor

16-12. In the circuit of Fig. 16-11, what is the critical frequency of the emitter bypass circuit?

16-13. In the circuit of Fig. 16-11, what would C_E be for the emitter bypass circuit to have a critical frequency of 20 Hz?

16-14. Find the critical frequency of the emitter bypass circuit in Fig. 16-12 when $\beta = 100$.

16-15. If the circuit of Fig. 16-11 had the following critical frequencies:

$f_c = 8$ Hz (input coupling circuit)
$f_c = 48$ Hz (output coupling circuit)
$f_c = 30$ Hz (emitter bypass circuit)

What is the lowest frequency in the midband of the amplifier?

16-16. If $\beta = 300$ in the circuit of Fig. 16-12, what is the lowest frequency in the midband of the amplifier?

16-17. If $\beta = 300$ in the circuit of Fig. 16-13, what is the lowest frequency in the midband of the amplifier?

Sec. 16-5 Collector Bypass Circuit

Note: In this section the effects of all the internal capacitances of the transistor will be neglected.

16-18. For the circuit of Fig. 16-11, find the critical frequency of the collector bypass circuit for the following values of C_{stray}:
  a. 10 pF   b. 20 pF

16-19. In the circuit of Fig. 16-11, what would C_{stray} be for the collector bypass circuit to have a critical frequency of 500 kHz?

16-20. Find the critical frequency of the collector bypass circuit in Fig. 16-12 when $\beta = 100$.

16-21. Find the critical frequency of the emitter bypass circuit in Fig. 16-13 when $\beta = 100$.

Sec. 16-6 Miller's Theorem

Note: In this section the effects of all the internal capacitances of the transistor will be neglected.

16-22. In the circuit of Fig. 16-14, $A = 100$ and $C = 10$ pF.
  a. Find $C_{in(Miller)}$.
  b. Find $C_{out(Miller)}$.

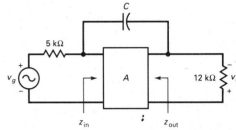

Figure 16-14

16-23. The circuit of Fig. 16-15 is an ac equivalent circuit of a CE amplifier that has an $r'_e = 30\ \Omega$. If a 10-pF capacitor is placed between the base and collector, determine the following:
  a. $C_{in(Miller)}$
  b. $C_{out(Miller)}$

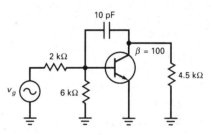

Figure 16-15

16-24. The input Miller capacitance in the circuit of Fig. 16-14 creates a bypass circuit on the input side. If $A = 100$, $C = 20$ pF, and $z_{in} = 5$ kΩ, what is the critical frequency of the bypass circuit?

16-25. The output Miller capacitance in the circuit of Fig. 16-14 creates a bypass circuit on the output side. If $A = 100$, $C = 20$ pF, and $z_o = 6$ kΩ, what is the critical frequency of the bypass circuit?

16-26. With respect to Prob. 16-23, what is the critical frequency of the input bypass circuit that is caused by the Miller effect?

16-27. With respect to Prob. 16-23, what is the critical frequency of the output bypass circuit that is caused by the Miller effect?

Sec. 16-7 High-Frequency Bipolar Analysis

Note: In this section consider the effects of the capacitance of the emitter and collector diodes of the transistor.

16-28. A transistor data sheet lists $f_T = 200$ MHz. What is the value of C'_e for $I_E = 5$ mA?

16-29. A transistor data sheet lists $C'_c = 2$ pF and the collector-base time constant $(r'_b C'_c)$ as 100 ps for $I_E = 10$ mA. What is r'_b for $I_E = 10$ mA?

16-30. The circuit of Fig. 16-16 is an ac equivalent circuit of a CE amplifier above midband. Determine the following:
  a. The critical frequency of the input bypass circuit.
  b. The critical frequency of the output bypass circuit.

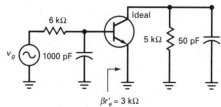

Figure 16-16

16-31. In the circuit of Fig. 16-17, $r'_b = 200\ \Omega$, $\beta = 200$, $r'_e = 10\ \Omega$, $C'_c = 4$ pF, $C'_e = 100$ pF, and $C_{stray} = 20$ pF.
Determine the following:
a. The total input bypass capacitance.
b. The total output bypass capacitance.
c. The critical frequency of the input bypass circuit.
d. The critical frequency of the output bypass circuit.
e. The dominant critical frequency above midband.

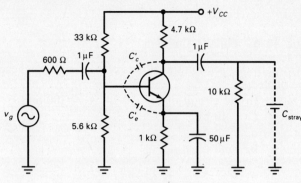

Figure 16-17

16-32. In the circuit of Fig. 16-18, $f_T = 100$ MHz, $r'_b C'_c = 300$ ps, $\beta = 100$, $C'_c = 3$ pF, and $C_{stray} = 15$ pF.
Determine the following:
a. r'_b
b. C'_e
c. The total input bypass capacitance.
d. The total output bypass capacitance.
e. The critical frequency of the input bypass circuit.
f. The critical frequency of the output bypass circuit.
g. The dominant critical frequency above midband.

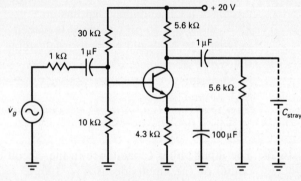

Figure 16-18

16-33. For the circuit of Fig. 16-18, if $f_T = 50$ MHz, $r'_b C'_c = 400$ ps, $\beta = 200$, $C'_c = 5$ pF, and $C_{stray} = 20$ pF, what is the dominant critical frequency above midband?

Sec. 16-8 The Total Frequency Response

16-34. An audio amplifier has a midband output voltage of 1 V. The lower critical frequency is 15 Hz and the upper critical frequency is 30 kHz. Determine the output voltage for the following frequencies:
    a. 5 Hz   b. 15 Hz   c. 30 Hz   d. 15 kHz   e. 30 kHz   f. 90 kHz

Sec. 16-9 Decibels

16-35. Calculate the decibel power gain G' of an amplifier that has the following ordinary power gains G:
    a. 2   b. 10   c. 20   d. 100   e. 500   f. 1000

16-36. An amplifier has an input power of 1 mW and an output power of 1 W.
Determine the following:
a. The ordinary power gain of the amplifier.
b. The power gain of the amplifier in decibels.

16-37. Calculate the ordinary power gain G of an amplifier that has the following decibel power gains G':
    a. 3 dB   b. 6 dB   c. 10 dB   d. 20 dB   e. 25 dB   f. 40 dB

Sec. 16-10 Decibel Voltage Gain

16-38. Calculate the decibel voltage gain A' of an amplifier that has the following ordinary voltage gains A:
   a. 2   **b.** 10   **c.** 20   **d.** 100   **e.** 500   **f.** 1000

16-39. An amplifier has an input voltage of 5 mV and an output voltage of 2 V. Determine the following:
   a. The ordinary voltage gain of the amplifier.
   b. The power gain of the amplifier in decibels.

16-40. Calculate the ordinary voltage gain A of an amplifier that has the following decibel voltage gains A':
   a. 3 dB   **b.** 6 dB   **c.** 10 dB   **d.** 20 dB   **e.** 25 dB   **f.** 40 dB

16-41. A two-stage amplifier has these stage gains: $A_1 = 50$ and $A_2 = 100$. Determine the following:
   a. The total ordinary voltage gain.
   b. The decibel voltage gain of each stage.
   c. The total decibel voltage gain.

16-42. A two-stage amplifier has these stage gains in decibels: $A'_1 = 20$ dB and $A'_2 = 30$ dB. Determine the following:
   a. The total decibel voltage gain.
   b. The ordinary voltage gain of each stage.
   c. The total ordinary voltage gain.

Sec. 16-11 Voltage Gain Outside the Midband

16-43. The two dominant critical frequencies of an amplifier are $f_1 = 200$ Hz and $f_2 = 200$ kHz. The decibel midband voltge gain is 50 dB. Sketch the ideal decibel graph of the frequency response.

16-44. The two dominant critical frequencies of an amplifier are $f_1 = 100$ Hz and $f_2 = 100$ kHz. The midband voltage gain is 100. Using five-cycle semilog paper, draw the ideal Bode plot (the decibel voltage gain versus frequency).

Name _____ Date _____

CHAPTER 17
OP-AMP THEORY

Study Chap. 17 in *Electronic Principles*.

I. TRUE / FALSE

Answer true (T) or false (F) to each of the following statements.

Answer

1. An operational amplifier (op amp) can only amplify ac signals. **(17-1)** 1. ___
2. Op amps are low-gain amplifiers. **(17-1)** 2. ___
3. A circuit where all the components are not soldered together or otherwise mechanically connected but produced and connected during the manufacturing process is called an integrated circuit (IC). **(17-1)** 3. ___
4. Monolithic ICs are limited to low-power applications. **(17-1)** 4. ___
5. Thin-film and thick-film ICs are made up of a combination of integrated and discrete components. **(17-1)** 5. ___
6. The kind of isolation that exists between the IC's integrated components and is caused by the depletion layers in the substrate is known as depletion-layer isolation. **(17-1)** 6. ___
7. Transistors, diodes, and resistors are the only practical components that can be produced on a chip. **(17-2)** 7. ___
8. Stages on a monolithic IC are capacitively coupled. **(17-2)** 8. ___
9. When the transistors are identical in a differential amplifier (diff amp), the emitter currents are equal. **(17-2)** 9. ___
10. The tail current in a diff amp is equal to the sum of the emitter currents. **(17-2)** 10. ___
11. A double-ended output diff amp with identical transistors does not have any potential difference between the collectors when the input voltages are the same. **(17-2)** 11. ___
12. The final stage of an op amp is typically a class A power amplifier. **(17-2)** 12. ___
13. The difference in base currents of a diff amp indicates how closely matched the transistors are. **(17-3)** 13. ___
14. Data sheets for an op amp always include the values of the base currents. **(17-3)** 14. ___
15. The voltage that a diff amp amplifies is the voltage difference between its two inputs. **(17-4)** 15. ___
16. The tail of a diff amp acts like a current source. **(17-4)** 16. ___

Copyright © 1993 by the Glencoe Division of Macmillan/McGraw-Hill School Publishing Company. All rights reserved.   167

17. The output voltage is ideal in a diff amp when both transistors are perfectly matched and inputs are grounded. (17-5)

18. Different values of V_{BE} are the only possible causes for an output offset voltage. (17-5)

19. The common-mode voltage gain of a diff amp is large. (17-6)

20. The diff amp discriminates between the desired input signal and the common-mode signal. (17-6)

II. COMPLETION

Complete each of the following statements.

Answer

1. The term operational amplifier refers to an amplifier that carries out ____ operations. (17-1)
2. A circuit where all the components have been soldered together or otherwise mechanically connected is called a ____ circuit. (17-1)
3. A wafer of p-type material that is to be used as the chassis for integrated components is called the p-____. (17-1)
4. Integrated circuits that have all their components as part of one chip are called ____ ICs. (17-1)
5. The most common type of integrated circuit is the ____ IC. (17-1)
6. Large-scale integration (LSI) refers to ICs with more than a ____ integrated components on a chip. (17-1)
7. The diff amp is widely used as the input stage of an ____ amplifier. (17-2)
8. Capacitors that have been fabricated on a chip are usually less than ____ pF. (17-2)
9. The current through the emitter resistor of a diff amp is called the ____ current. (17-2)
10. If you include the voltage drop across the emitter diode in a diff amp, the value of the node voltage at the top of the tail resistor is closest to ____ V. (17-2)
11. In a single-ended output diff amp, the input voltage that is in phase with the output voltage is called the ____ input. (17-2)
12. In a single-ended output diff amp, the input voltage that is 180° out of phase with the output voltage is called the ____ input. (17-2)
13. The difference between the base currents in a diff amp is defined as the input ____ current. (17-3)
14. The average of the two base currents in a diff amp is defined as the input ____ current. (17-3)
15. The voltage gain of a single-ended output diff amp is equal to the collector resistance divided by ____. (17-4)
16. The input impedance of a diff amp is equal to β times ____. (17-4)
17. The smallest differences between the two transistors of a diff amp are amplified to produce an output ____ voltage. (17-5)
18. An output offset voltage can be nulled or eliminated by applying a small dc ____ voltage. (17-5)
19. The CMRR of a diff amp stands for the common-mode ____ ratio. (17-6)
20. The CMRR is defined as the ratio of the ____ voltage gain to the common-mode voltage gain. (17-6)

III. ILLUSTRATIVE AND PRACTICE PROBLEMS

Illustrative Problem 1. For the diff amp with identical transistors and $\beta_{dc} = 100$, find the following:

a. I_T
b. I_E
c. V_{out}

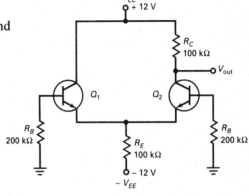

Figure 17-1

Steps	Comments
1. Find I_T. Figure 17-2 $I_T = \dfrac{12\text{ V} - 0.7\text{ V}}{100\text{ k}\Omega} = 113\ \mu\text{A}$	Since the transistors are identical, the exact equation for I_T can be derived by applying Kirchhoff's voltage law around the closed loop. $I_T = \dfrac{V_{EE} - 0.7\text{ V}}{R_E - \dfrac{R_B}{2\beta_{dc}}}$ You can neglect the term $R_B/2\beta_{dc}$ from the above equation when $R_E > 20\ R_B/2\beta_{dc}$ (second approximation). The voltage drop across R_B equals $I_B R_B$. Since $I_B = I_C/\beta_{dc}$, $I_C \approx I_E$, and $I_E = I_T/2$, the voltage drop across R_B can be shown to be $I_T R_B/2\beta_{dc}$ by direct substitution. $R_E = 100\text{ k}\Omega \qquad \dfrac{R_B}{2\beta_{dc}} = 1\text{ k}\Omega$ Since $R_E > 20\ \dfrac{R_B}{2\beta_{dc}}$, neglect $\dfrac{R_B}{2\beta_{dc}}$
2. Find I_E. $I_E = \dfrac{113\ \mu\text{A}}{2} = 56.5\ \mu\text{A}$	Since the transistors are identical, $I_E = \dfrac{I_T}{2}$
3. Find V_{out}. $V_{out} = 12\text{ V} - (56.5\ \mu\text{A})(100\text{ k}\Omega) = 6.35\text{ V}$	Output voltage. $V_{out} = V_{CC} - I_C R_C \qquad I_C \approx I_E$ $V_{out} = V_{CC} - I_E R_C \qquad$ by substitution

Practice Problem 1. For the diff amp with identical transistors and $\beta_{dc} = 200$, find the following:

a. I_T
b. I_E
c. V_{out}

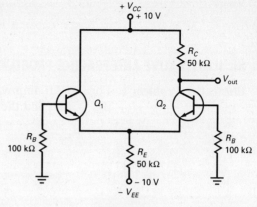

Figure 17-3

Answers: **a.** $I_T = 0.186$ mA **b.** $I_E = 0.093$ mA **c.** $V_{out} = 5.35$ V

Illustrative Problem 2. For the diff amp with identical transistors and $\beta_{dc} = \beta = 100$, find the following:

a. I_T
b. I_E
c. r'_e
d. A_{CM}
e. $v_{out(CM)}$

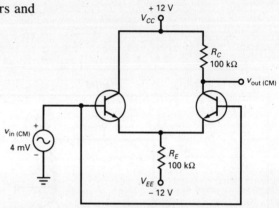

Figure 17-4

Steps	Comments
1. Find I_T. $$I_T = \frac{12\text{ V} - 0.7\text{ V}}{100\text{ k}\Omega} = 113\ \mu\text{A}$$	See Illus. Prob. 1. $$I_T = \frac{V_{EE} - 0.7\text{ V}}{R_E}$$
2. Find I_E. $$I_E = \frac{113\ \mu\text{A}}{2} = 56.5\ \mu\text{A}$$	See Illus. Prob. 1. $$I_E = \frac{I_T}{2}$$
3. Find r'_e. $$r'_e = \frac{25\text{ mV}}{56.5\ \mu\text{A}} = 442\ \Omega$$	Transistor dynamic resistance. $$r'_e = \frac{25\text{ mV}}{I_E}$$

Steps	Comments
4. Find A_{CM}. 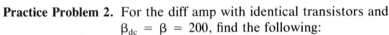 Figure 17-5 $$A_{CM} = \frac{100 \text{ k}\Omega}{2(100 \text{ k}\Omega)} = 0.5$$	**a.** Draw the ac equivalent circuit for a common-mode input. **b.** The common-mode voltage gain. $$A_{CM} = \frac{v_{out(CM)}}{v_{in(CM)}} = \frac{i_c R_C}{i_e(r'_e + 2R_E)} = \frac{R_C}{r'_e + 2R_E}$$ $$i_c \approx i_e$$ $$A_{CM} = \frac{R_C}{2R_E} \quad \text{since } 2R_E \gg r'_e$$
5. Find $v_{out(CM)}$. $v_{out(CM)} = (0.5)(4 \text{ mV}) = 2 \text{ mV}$	The common-mode output voltage. $v_{out(CM)} = A_{CM} v_{in(CM)}$

Practice Problem 2. For the diff amp with identical transistors and $\beta_{dc} = \beta = 200$, find the following:

a. I_T
b. I_E
c. r'_e
d. A_{CM}
e. $v_{out(CM)}$

Answers: **a.** $I_T = 0.186$ mA **b.** $I_E = 93$ μA **c.** $r'_e = 269 \, \Omega$
d. $A_{CM} = 0.5$ **e.** $v_{out(CM)} = 1$ mV

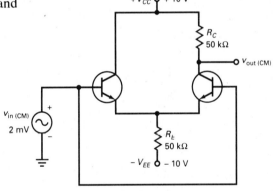

Figure 17-6

Illustrative Problem 3. For the diff amp with identical transistors and $\beta_{dc} = \beta = 100$, find the following:

a. v_{in} (differential input voltage)
b. A (differential voltage gain)
c. z_{in} (differential input impedance)
d. v_{out} (differential output voltage)
e. CMRR

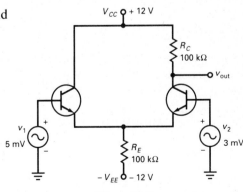

Figure 17-7

Steps	Comments
1. Find v_{in}. $v_{in} = 5\text{ mV} - 3\text{ mV} = 2\text{ mV}$	The differential input voltage. $v_{in} = v_1 - v_2$
2. Find A. Figure 17-8 $A = \dfrac{100\text{ k}\Omega}{2(442\ \Omega)} = 113$	a. Draw the ac equivalent circuit for a differential input. b. The dc biasing circuit is the same as the dc biasing circuit in Illus. Prob. 2. See Illus. Prob. 2. $r'_e = 442\ \Omega \qquad A_{CM} = 0.5$ c. The differential voltage gain. $A = \dfrac{v_{out}}{v_{in}} = \dfrac{i_c R_C}{i_e(2r'_e)} = \dfrac{R_C}{2r'_e}$ $i_c \approx i_e$
3. Find z_{in}. $z_{in} = 2(100)(442\ \Omega) = 88.4\text{ k}\Omega$	Differential input impedance. $z_{in} = \dfrac{v_{in}}{i_b} = \dfrac{i_e(2r'_e)}{i_b} = 2\beta r'_e$ $\dfrac{i_e}{i_b} \approx \beta$
4. Find v_{out}. $v_{out} = 113(2\text{ mV}) = 226\text{ mV}$	Differential output voltage. $v_{out} = A v_{in}$
5. Find CMRR. $\text{CMRR} = \dfrac{113}{0.5} = 226$	Common-mode rejection ratio. $\text{CMRR} = \dfrac{A}{A_{CM}}$

Practice Problem 3. For the diff amp with identical transistors and $\beta_{dc} = \beta = 200$, find the following:
a. v_{in} (differential input voltage)
b. A (differential voltage gain)
c. z_{in} (differential input impedance)
d. v_{out} (differential output voltage)
e. CMRR

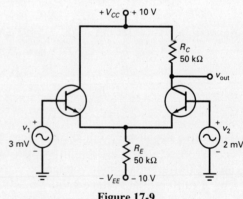

Figure 17-9

Answers: a. $v_{in} = 1\text{ mV}$ b. $A = 92.9$ c. $z_{in} = 108\text{ k}\Omega$ d. $v_{out} = 92.9\text{ mV}$
e. CMRR = 186

Name _____ Date _____

IV. PROBLEMS

Note: For this chapter, make the following assumptions:

1. The transistors in all the diff amps are identical unless specified otherwise.
2. $\beta_{dc} = \beta = 100$ for all the transistors unless specified otherwise.
3. Use the second approximation of I_T unless specified otherwise.

Sec. 17-2 The Differential Amplifier

17-1. For the circuit of Fig. 17-10, find the following:
  a. I_T
  b. I_E
  c. V_{out}

17-2. Repeat Prob. 17-1, using the ideal approximation of I_T.

17-3. Repeat Prob. 17-1, using the exact equation for I_T.

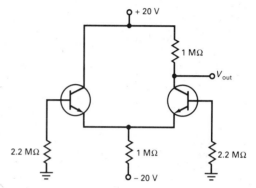

Figure 17-10

17-4. Repeat Prob. 17-1 for the circuit of Fig. 17-11.

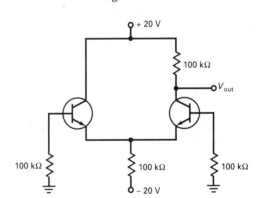

Figure 17-11

17-5. What is the output voltage for the circuit of Fig. 17-12?

Sec. 17-3 Two Input Characteristics

17-6. If the base currents of a diff amp are 75 nA and 65 nA, determine the following:
  a. Input offset current.
  b. Input bias current.

17-7. A data sheet gives an input bias current of 50 nA and an input offset current of 6 nA. What are the base currents?

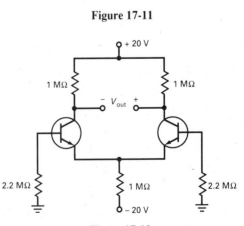

Figure 17-12

Copyright © 1993 by the Glencoe Division of Macmillan/McGraw-Hill School Publishing Company. All rights reserved.

Sec. 17-4 AC Analysis of a Diff Amp

17-8. For the diff amp in Fig. 17-13 with a tail current of 30 μA, determine the following:
  a. A
  b. z_{in}
  c. v_{out}

17-9. If the diff amp of Fig. 17-13 had a tail current of 20 μA, determine the following:
  a. A
  b. z_{in}
  c. v_{out}

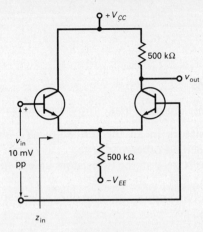

Figure 17-13

17-10. For the diff amp of Fig. 17-14, if the reading of the dc voltmeter is −0.7 V, determine the following:
  a. A
  b. z_{in}
  c. v_{out}

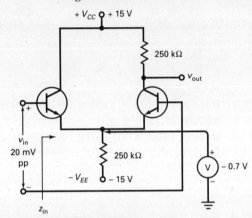

17-11. If the diff amp of Fig. 17-13 had a tail current of 50 μA, find V_{out}.

Sec. 17-6 Common Mode Gain

17-12. For the circuit of Fig. 17-15, determine the following:
  a. A_{CM}
  b. $v_{out(CM)}$

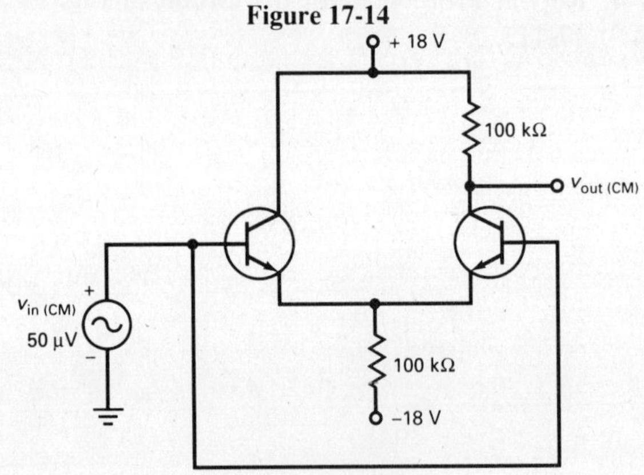

Figure 17-14

17-13. In the circuit of Fig. 17-13, if $I_T = 40$ μA, $v_{in} = 6$ mV, and a common-mode voltage $v_{in(CM)}$ of 4 mV exists on both bases, find the following:
  a. A e. $v_{out(total)}$
  b. A_{CM} f. CMRR
  c. v_{out} g. CMRR′
  d. $v_{out(CM)}$

17-14. For the circuit of Fig. 17-16 with $v_{in} = 10$ mV and $v_{in(CM)} = 2$ mV, determine the following:
  a. A
  b. A_{CM}
  c. v_{out}
  d. $v_{out(CM)}$
  e. $v_{out(total)}$
  f. CMRR
  g. CMRR′

Figure 17-15

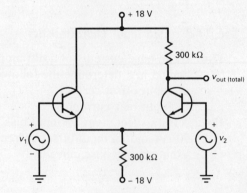

17-15. What is the CMRR of the diff amp in Prob. 17-10?

17-16. The data of an op amp gives $A = 200,000$ and CMRR′ = 100 dB. What is the common-mode voltage gain?

Figure 17-16

CHAPTER 18

MORE OP-AMP THEORY

Study Chap. 18 in *Electronic Principles*.

I. TRUE / FALSE

Answer true (T) or false (F) to each of the following statements.

Answer

1. The input stage of a typical op amp is a diff amp. **(18-1)** 1. ___
2. The frequency at which the voltage gain of the op amp becomes 1 is called the unity-gain frequency. **(18-1)** 2. ___
3. A compensating capacitor prevents unwanted oscillation. **(18-1)** 3. ___
4. In a typical op amp, f_{unity} is not much greater than f_c. **(18-1)** 4. ___
5. The initial slope of the exponential charging of the compensating capacitor in a diff amp refers to the slew rate of the amplifier. **(18-2)** 5. ___
6. Slew-rate distortion occurs when the slew rate of the amplifier is greater than the initial slope of the output signal. **(18-2)** 6. ___
7. Slew-rate distortion makes a sinusoidal output signal look more triangular. **(18-2)** 7. ___
8. If the compensating capacitor of a diff amp is increased, the slew rate of the amplifier would increase. **(18-2)** 8. ___
9. The units for the slew rate are V/μs. **(18-2)** 9. ___
10. If the peak value of the output voltage of an op amp is increased, then its power bandwidth would decrease. **(18-3)** 10. ___
11. If the slew rate of an op amp is increased, then its power bandwidth would decrease. **(18-3)** 11. ___
12. If the bases of the transistors in a diff amp do not have a dc path to ground, the transistors will go into saturation. **(18-4)** 12. ___
13. The least expensive and most widely used op amp is the 741C. **(18-4)** 13. ___
14. The input stage of an LF13741 BIFET op amp is a diff amp. **(18-4)** 14. ___
15. As long as an op amp is operating in its linear region, the output of the op amp is equivalent to a Thevenin circuit. **(18-4)** 15. ___
16. An op amp output offset voltage can be caused by an input offset voltage. **(18-5)** 16. ___
17. An op amp output offset voltage can be caused by the input transistors having different β_{dc}'s. **(18-5)** 17. ___
18. If the frequency of a 741C op amp is increased, its CMRR' will increase. **(18-5)** 18. ___

19. The slew rate limits the high-frequency large-signal response of an op amp. (18-5) 19. ___
20. The BIFET op amp's input bias and offset currents are much smaller than the bipolar op amp input bias and offset currents. (18-6) 20. ___

II. COMPLETION

Complete each of the following statements.

Answer

1. Under certain conditions, a high-gain amplifier can produce an unwanted high-frequency signal called an ___. (18-1) 1. ___
2. The device that allows the designer to control the dominant critical frequency of an op amp is the ___ capacitor. (18-1) 2. ___
3. At f_{unity}, the voltage gain of an op amp is ___. (18-1) 3. ___
4. When a diff amp is operating beyond its critical frequency, its voltage gain decreases at a rate of ___ dB per decade. (18-1) 4. ___
5. If the frequency of a sine wave is increased, its initial slope will ___. (18-2) 5. ___
6. If the peak value of a sine wave is increased, its initial slope will ___. (18-2) 6. ___
7. The fastest response an amplifier is capable of is limited by its ___ rate. (18-2) 7. ___
8. If the maximum charging current of the compensating capacitor in a diff amp were to increase, then the slew rate of the amplifier would ___. (18-2) 8. ___
9. The maximum rate of output voltage change is called the ___ rate. (18-2) 9. ___
10. The maximum frequency without slew-rate distortion is sometimes called the ___ bandwidth of the op amp. (18-3) 10. ___
11. The critical condition in an op amp that separates normal operation from slew-rate distortion is when S_R is equal to ___. (18-3) 11. ___
12. The type of loading that uses transistors instead of resistors is called ___ loading. (18-4) 12. ___
13. The typical input impedance of the 741C op amp is ___ Ω. (18-4) 13. ___
14. The op amp with an extremely high input impedance that combines FETs and bipolar transistors is called a ___ op amp. (18-4) 14. ___
15. The typical voltage gain of the 741C op amp is ___. (18-4) 15. ___
16. The typical output impedance of the 741C op amp is ___ Ω. (18-4) 16. ___
17. An input offset voltage of an op amp is caused when the input transistors have different ___ curves. (18-5) 17. ___
18. The data sheet of a typical 741C op amp lists a maximum input offset voltage of ± ___ mV. (18-5) 18. ___
19. At low frequencies the CMRR' of the 741C op amp is ___ dB. (18-5) 19. ___
20. An op amp output offset voltage can be produced by the input transistors having different ___ resistors. (18-5) 20. ___
21. The LM318 op amp has an extremely high slew rate of ___. (18-6) 21. ___

III. ILLUSTRATIVE AND PRACTICE PROBLEMS

Illustrative Problem 1. An op amp has $A = 200{,}000$ and $S_R = 1$ V/μs. If the op amp is operating at a frequency of 50 kHz and $v_{in} = 40$ μV, determine the following:

a. v_{out}
b. V_P
c. f_{max}
d. Is slew-rate distortion present?

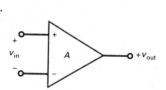

Figure 18-1

Steps	Comments
1. Find v_{out}. $v_{out} = 200{,}000(40\ \mu V) = 8\ V$	$v_{out} = A v_{in}$
2. Find V_P. $V_P = \dfrac{8\ V}{2} = 4\ V$	Peak value of the output voltage. $V_P = \dfrac{v_{out}}{2}$
3. Find f_{max}. $f_{max} = \dfrac{1\ V/\mu s}{2\pi(4\ V)} = 39.8\ kHz$	Power bandwidth—the maximum operating frequency without slew-rate distortion. $f_{max} = \dfrac{S_R}{2\pi V_P}$ $f = 50\ kHz$ $f > f_{max}$ Slew-rate distortion is present because the operating frequency is greater than the power bandwidth. The output voltage waveform is distorted.

Practice Problem 1. An op amp has $A = 100{,}000$ and $S_R = 0.5$ V/μs. If the op amp is operating at a frequency of 20 kHz and $v_{in} = 100$ μV, determine the following:

a. v_{out}
b. V_P
c. f_{max}
d. Is slew-rate distortion present?

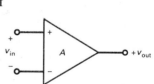

Figure 18-2

Answers: **a.** $v_{out} = 10$ V **b.** $V_P = 5$ V **c.** $f_{max} = 15.9$ kHz **d.** Yes

IV. PROBLEMS

Sec. 18-1 Small-Signal Frequency Response

18-1. For the circuit of Fig. 18-3, if $A_{mid} = 100$, determine the following:
 a. f_c
 b. f_{unity}

18-2. For the circuit of Fig. 18-3, if $I_T = 0.25$ mA and the transistors are identical, determine the following:
 a. f_c
 b. f_{unity}

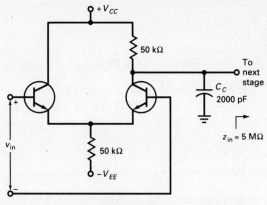

Figure 18-3

18-3. The data sheet of an op amp gives $A_{mid} = 200{,}000$ and $f_{unity} = 5$ MHz. What is the critical frequency of the op amp?

18-4. An op amp has a critical frequency of 25 Hz and a unity-gain frequency of 4 MHz. What is its midband voltage gain?

18-5. An op amp has a critical frequency of 30 Hz and a midband voltage gain of 200,000. What is its unity-gain frequency?

Sec. 18-2 Large-Signal Frequency Response

18-6. For the sinusoidal voltage waveform of Fig. 18-4, do the following:
 a. Complete Table 18-1 for the given sine wave conditions.

V_P V	f kHz	S_S V/μs
10	20	
20	20	
10	40	

Table 18-1

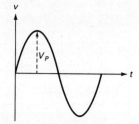

Figure 18-4

Select either *increased*, *decreased*, or *stayed the same* as the correct answer to the following questions.
 b. What happened to S_S when only V_P was increased?
 c. What happened to S_S when only f was increased?

18-7. To get a slew rate of 2 V/μs with a compensating capacitor of 1000 pF, what should the maximum charging current be?

18-8. If a compensating capacitor of 2000 pF has a maximum charging current of 2 mA, what is the slew rate?

18-9. To get a slew rate of 10 V/μs with a maximum charging current of 100 μA, what value should the compensating capacitor be?

18-10. An LM307 has a slew rate of 0.5 V/μs. Will slew-rate distortion occur if the frequency the op amp is operating at is 50 kHz and the peak value of its sinusoidal output voltage is 10 V?

18-11. An LM318 has a slew rate of 70 V/μs. Will slew-rate distortion occur if the frequency the op amp is operating at is 1 MHz and the peak-to-peak value of its sinusoidal output voltage is 12 V?

18-12. An op amp is operating at a frequency of 100 kHZ and the peak value of its sinusoidal output voltage is 2 V. What is the minimum slew rate that the op amp can have to prevent slew-rate distortion from occurring?

Sec. 18-3 Power Bandwidth

18-13. For the op amp of Fig. 18-5, do the following:
   a. Complete Table 18-2 for the given circuit conditions.

V_P V	S_R V/μs	f_{max} kHz
5	1	
10	1	
5	2	

Table 18-2

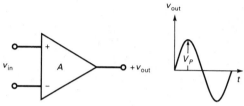

Figure 18-5

Select either *increased*, *decreased*, or *stayed the same* as the correct answer to the following questions.

b. What happened to f_{max} when only V_P was increased?
c. What happened to f_{max} when only S_R was increased?

18-14. An op amp has a slew rate of 10 V/μs and a peak output voltage of 5 V. What is the maximum frequency the op amp can operate at without having slew-rate distortion?

18-15. An op amp is operating at 50 kHz and has a slew rate of 0.5 V/μs. What is the largest undistorted peak-to-peak output voltage the op amp can have?

Sec. 18-4 The Operational Amplifier

18-16. If the op amp in Fig. 18-5 has a voltage gain of 200,000 and an input voltage of 20 μV, what is its output voltage?

18-17. If the op amp in Fig. 18-5 has a voltage gain of 400,000 and an output voltage of 4 V, what is its input voltage?

CHAPTER 19

OP-AMP NEGATIVE FEEDBACK

Study Chap. 19 in *Electronic Principles*.

I. TRUE / FALSE

Answer true (T) or false (F) to each of the following statements.

Answer

1. An op amp with noninverting voltage feedback tends to act like an ideal voltage amplifier. **(19-1)**   1. ___
2. An ideal voltage amplifier is described as having an infinite input impedance, a zero output impedance, and a constant voltage gain. **(19-1)**   2. ___
3. With noninverting voltage feedback, any increase in the differential voltage gain of the op amp produces an increase in error voltage. **(19-1)**   3. ___
4. With noninverting voltage feedback, a large change in the differential voltage gain of an op amp produces a significant change in the output voltage. **(19-1)**   4. ___
5. The error voltage of an op-amp with noninverting voltage feedback is very small. **(19-1)**   5. ___
6. The open-loop voltage gain of an op amp is much larger than its closed-loop voltage gain. **(19-2)**   6. ___
7. The output voltage of an op amp with noninverting voltage feedback is equal to the differential input voltage times the closed-loop gain. **(19-2)**   7. ___
8. When the loop gain of an op amp with noninverting voltage feedback is much greater than 1, then the closed-loop voltage gain equals $1/B$. **(19-2)**   8. ___
9. With noninverting voltage feedback, the closed-loop input impedance of an op amp is larger than its open-loop input impedance. **(19-3)**   9. ___
10. With noninverting voltage feedback, the closed-loop output impedance of an op amp is larger than its open-loop output impedance. **(19-3)**   10. ___
11. Nonlinear distortion occurs in the final stage of an op amp when large swings in current cause the r'_e of the transistor to change during the cycle. **(19-4)**   11. ___
12. Noninverting voltage feedback reduces nonlinear distortion by making the closed-loop voltage gain independent of changes in the open-loop voltage gain. **(19-4)**   12. ___
13. If the loop gain of an op amp with noninverting voltage feedback is increased, then its nonlinear distortion will decrease. **(19-4)**   13. ___
14. With inverting voltage feedback, an op amp acts like an ideal current-to-voltage converter. **(19-5)**   14. ___
15. An ideal current-to-voltage converter is a device with zero input impedance, zero output impedance, and constant transresistance. **(19-5)**   15. ___

16. The transresistance is defined as the ratio of the output voltage to the input current. (19-5)

16. ____

17. With inverting voltage feedback, if the feedback resistor of an op amp is increased, then its output voltage will decrease. (19-5)

17. ____

18. The open-loop gain-bandwidth product of an op amp is larger than its closed-loop gain-bandwidth product. (19-6)

18. ____

19. Negative feedback has no effect on the slew rate and the power bandwidth. (19-6)

19. ____

II. COMPLETION

Complete each of the following statements.

Answer

1. The differential input of an op amp with noninverting voltage feedback is called the ____ voltage. (19-1)

1. ____

2. The product AB is called the ____ gain. (19-1)

2. ____

3. In an op amp with noninverting voltage feedback, B represents the fraction of the output voltage fed back to the input through a voltage ____. (19-1)

3. ____

4. In order for the voltage gain of an op amp with noninverting voltage feedback to solely depend on the feedback resistors, the product of AB must be much greater than ____. (19-1)

4. ____

5. The ideal op amp differential voltage gain is ____. (19-1)

5. ____

6. When there are no loading effects, the open-loop voltage gain of an op amp is equal to its ____ voltage gain. (19-2)

6. ____

7. If the voltage gain of the op amp's voltage divider is increased, then the op amp's closed-loop voltage gain will ____. (19-2)

7. ____

8. If the voltage gain of the op amp's voltage divider is increased, then the op amp's output voltage will ____. (19-2)

8. ____

9. If the loop gain of an op amp with noninverting voltage feedback were increased, then its closed-loop input impedance would ____. (19-3)

9. ____

10. If the loop gain of an op amp with noninverting voltage feedback were increased, then its closed-loop output impedance would ____. (19-3)

10. ____

11. Because the quantity $1 + AB$ indicates how much the voltage is reduced by negative feedback, it is called the ____ of the feedback amplifier. (19-4)

11. ____

12. The quantity $1 + AB$ is also known as the ____ factor. (19-4)

12. ____

13. The type of distortion that occurs because the ac signal swings over most of the ac load line is called ____ distortion. (19-4)

13. ____

14. If the loop gain of an op amp with noninverting voltage feedback is increased, then its output offset voltage will ____. (19-4)

14. ____

15. A node that has 0 V with respect to ground without being mechanically grounded is called a ____ ground. (19-5)

15. ____

16. The inverting terminal of an ideal op amp with inverting voltage feedback is at ____ ground. (19-5)

16. ____

17. The input impedance of an ideal op amp with inverting voltage feedback is ____. (19-5)

17. ____

18. The closed-loop bandwidth of an op amp equals its open-loop bandwidth times the ____ factor. (19-6)

18. ____

19. If the closed-loop voltage gain of an op amp is decreased, then its closed-loop bandwidth will ____. (19-6)

19. ____

III. ILLUSTRATIVE AND PRACTICE PROBLEMS

Illustrative Problem 1. If the op amp in Fig. 19-1 has $A = 100{,}000$ and $f_{unity} = 5$ MHz, determine the following:

a. AB
b. A_{CL}
c. v_{out}
d. v_{error}
e. $f_{2(CL)}$

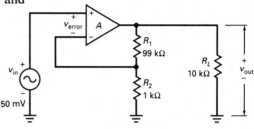

Figure 19-1

Steps	Comments
1. Find B. $B = \dfrac{1\text{ k}\Omega}{99\text{ k}\Omega + 1\text{ k}\Omega} = 0.01$	B is the fraction of the output voltage fed back to the input of the op amp. $B = \dfrac{R_2}{R_1 + R_2}$
2. Find AB. $AB = 100{,}000\,(0.01) = 1000$	Loop gain of the op amp circuit.
3. Find A_{CL}. $A_{CL} = \dfrac{1}{0.01} = 100$	Closed-loop gain of the op amp circuit. $A_{CL} = \dfrac{A}{1 + AB}$ $A_{CL} = \dfrac{1}{B}$ when $AB \gg 1$
4. Find v_{out}. $v_{out} = 100(50\text{ mV}) = 5\text{ V}$	Output voltage. $v_{out} = A_{CL} v_{in}$
5. Find v_{error}. $v_{error} = \dfrac{5\text{ V}}{100{,}000} = 50\ \mu\text{V}$	The error voltage is the differential input to the op amp. It should be very small. $v_{error} = \dfrac{v_{out}}{A}$
6. Find $f_{c(CL)}$. $f_{c(CL)} = \dfrac{5\text{ MHz}}{100} = 50\text{ kHz}$	Bandwidth of the op amp circuit. $f_{c(CL)} = \dfrac{f_{unity}}{A_{CL}}$

Practice Problem 1. If the op amp in Fig. 19-2 has $A = 100{,}000$ and $f_{unity} = 5$ MHz, determine the following:

a. AB
b. A_{CL}
c. v_{out}
d. v_{error}
e. $f_{2(CL)}$

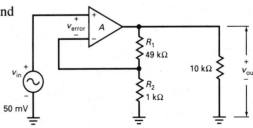

Figure 19-2

Answers: a. $AB = 2000$ b. $A_{CL} = 50$ c. $v_{out} = 2.5$ V
d. $v_{error} = 25\ \mu$V   e. $f_{2(CL)} = 100$ kHz

IV. PROBLEMS

Sec. 19-1 Noninverting Voltage Feedback

19-1. For the op amp circuit of Fig. 19-3, with $A = 500{,}000$, determine the following:
  a. B
  b. AB
  c. v_{out}
  d. v_{error}

19-2. The circuit of Fig. 19-4 has $A = 400{,}000$. Use $v_{out}/v_{in} = 1/B$ to determine the following:
  a. v_{out}
  b. v_{error}
  c. v_2
  d. How does v_2 compare to v_{in}?

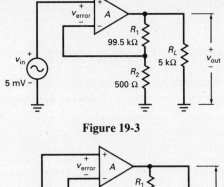

Figure 19-3

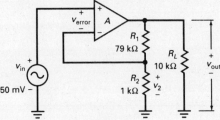

Figure 19-4

19-3. The circuit of Fig. 19-4 has $A = 400{,}000$. Use $v_{out}/v_{in} = A/(1 + AB)$ to determine the following:
  a. v_{out}
  b. v_{error}
  c. v_2
  d. How does v_2 compare to v_{in}?

19-4. For the circuit of Fig. 19-5, do the following:
  a. Complete Table 19-1 for the given circuit conditions.

R_2, kΩ	R_L, kΩ	A	v_{out} V	v_{error} μV
1	10	100,000		
3	10	100,000		
1	20	100,000		
1	10	200,000		

Table 19-1

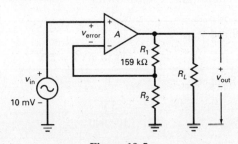

Figure 19-5

Select either *increased*, *decreased*, or *stayed the same* as the correct answer to the following questions.
  b. What happened to v_{out} when only R_2 was increased?
  c. What happened to v_{out} when only R_L was increased?
  d. What happened to v_{out} when only A was increased?
  e. What happened to v_{error} when only R_2 was increased?
  f. What happened to v_{error} when only R_L was increased?
  g. What happened to v_{error} when only A was increased?

19-5. For the op-amp circuit of Fig. 19-6, with $A = 400{,}000$, determine the following:
  a. v_{out}
  b. v_{error}

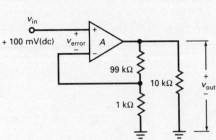

Figure 19-6

Sec. 19-2 Open-Loop and Closed-Loop Voltage Gains

19-6. The op amp of Fig. 19-3 has $z_{out} = 100\ \Omega$ and $A = 500{,}000$. If $v_{in} = 10$ mV, determine the following:
   a. A_{OL} b. A_{CL} c. v_{out} d. v_{error}

19-7. The op amp of Fig. 19-4 has $z_{out} = 50\ \Omega$ and $A = 400{,}000$. If $v_{in} = 10$ mV, determine the following:
   a. A_{OL} b. A_{CL} c. v_{out} d. v_{error}

Sec. 19-3 Input and Output Impedance

19-8. A circuit with an equivalent op amp is shown in Fig. 19-7. If $A = 200{,}000$, determine the following:
   a. $z_{in(CL)}$
   b. $z_{out(CL)}$

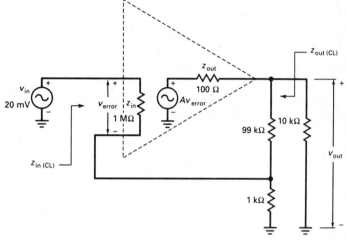

Figure 19-7

19-9. If the circuit of Fig. 19-3 has $A = 500{,}000$, $z_{in} = 1\ M\Omega$, and $z_{out} = 80\ \Omega$, determine the following:
   a. $z_{in(CL)}$
   b. $z_{out(CL)}$

Sec. 19-5 Inverting Voltage Feedback

19-10. For the circuit of Fig. 19-8, determine the following:
   a. v_{out}
   b. The current through the feedback resistor.
   c. The current through the load resistor.

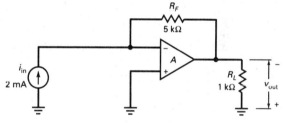

Figure 19-8

19-11. The op amp in Fig. 19-9 has $z_{in} = 2\ M\Omega$ and $z_{out} = 80\ \Omega$. If $A = 100{,}000$, determine the following:
   a. v_{out}
   b. v_{error}
   c. $z_{in(CL)}$
   d. $z_{out(CL)}$

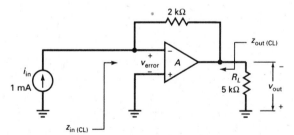

Figure 19-9

19-12. If the circuit of Fig. 19-10 has i_{in} = 5 mA, R_L = 2 kΩ, and R_F = 1 kΩ, what is the value of the output voltage?

19-13. If the circuit of Fig. 19-10 has i_{in} = 5 mA, v_{out} = 6 V, and R_L = 1 kΩ, what is the value of the feedback resistor?

19-14. If the circuit of Fig. 19-10 has v_{out} = 8 V, R_L = 20 kΩ, and R_F = 10 kΩ, what is the value of the input current?

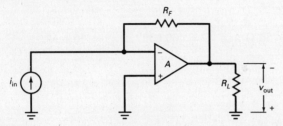

Figure 19-10

Sec. 19-6 Bandwidth

19-15. An op amp has A = 100,000 and f_{unity} = 4 MHz. If the closed-loop voltage gain is 200, determine the following:
   a. The open-loop bandwidth.
   b. The closed-loop bandwidth.

19-16. An op amp has f_{unity} = 5 MHz and f_2 = 20 Hz. If A_{CL} = 100, determine the following:
   a. $f_{2(CL)}$
   b. Open-loop voltage gain.

19-17. If the op amp of Fig. 19-3 has A = 200,000 and f_2 = 10 Hz, determine the following:
   a. f_{unity}
   b. $f_{2(CL)}$
   c. The gain-bandwidth product.

19-18. For the frequency response of an op amp in Fig. 19-11, determine the following:
   a. The open-loop voltage gain.
   b. f_{unity}

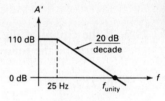

Figure 19-11

19-19. For the circuit and frequency response of the op amp in Fig. 19-12, determine the following:
   a. A
   b. f_2
   c. $f_{2(CL)}$
   d. The gain-bandwidth product.

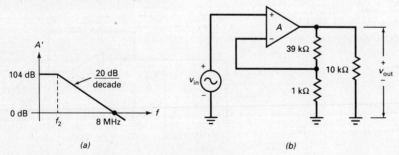

Figure 19-12

19-20. If the slew-rate of the op amp in Prob. 19-19 is 5 V/μs, what is maximum undistorted peak output voltage at $f_{2(CL)}$?

CHAPTER 20
LINEAR OP-AMP CIRCUITS

Study Chap. 20 in *Electronic Principles*.

I. TRUE / FALSE

Answer true (T) or false (F) to each of the following statements.

Answer

1. An amplified output from a linear op-amp circuit preserves the shape of the input signal. **(20-1)** 1. ___
2. An approximation that is worth memorizing is, if the op amp is not saturated, its two input voltages are equal. **(20-1)** 2. ___
3. To use an op amp you need either a dual or split supply. **(20-1)** 3. ___
4. If the source or the load has a dc voltage with respect to ground and the op amp is used as an ac amplifier, a coupling capacitor will prevent this dc voltage from interfering with the operation of the amplifier. **(20-1)** 4. ___
5. The virtual ground of an op-amp circuit is the same as an ac ground. **(20-2)** 5. ___
6. The input current of an inverting voltage amplifier (IVA) equals the input voltage divided by the series resistor. **(20-2)** 6. ___
7. The input current of an IVA flows through both the series and feedback resistors. **(20-2)** 7. ___
8. The output voltage of an IVA equals the open-loop voltage gain times the input voltage. **(20-2)** 8. ___
9. When a variable resistor is put across the input of the op amp in an IVA, the bandwidth of the amplifier can be adjusted without changing its closed-loop voltage gain. **(20-3)** 9. ___
10. The voltage gain for each input voltage in a summing amplifier is equal to 1. **(20-4)** 10. ___
11. In a summing amplifier, all the input series resistors are equal. **(20-4)** 11. ___
12. One way to obtain bidirectional load current from a current booster is to do so when the current booster is an emitter follower. **(20-5)** 12. ___
13. A floating load is a load that is not connected to ground. **(20-6)** 13. ___
14. An essential characteristic of a good instrumentation amplifier is that it is an extremely low noise device. **(20-7)** 14. ___
15. An instrumentation amplifier is typically used in applications in which a small difference voltage and a large common-mode voltage are the inputs. **(20-7)** 15. ___
16. A low-pass filter will block direct current. **(20-8)** 16. ___

17. The Butterworth filter is a passive filter. (20-8) 17. ___
18. A pole can be a coupling circuit that appears in an active filter. (20-8) 18. ___
19. At the cutoff frequency in a two-pole Butterworth filter, the overall voltage gain is down 6 dB. (20-8) 19. ___
20. Above the cutoff frequency of a one-pole low-pass filter, the output decreases at a rate of 20 dB per decade. (20-8) 20. ___

II. COMPLETION

Complete each of the following statements.

Answer

1. Op-amp circuits that never saturate under normal operating conditions are called ___ op-amp circuits. (20-1) 1. _____
2. Sometimes a bypass capacitor with a low critical frequency is put in an amplifier to prevent ___ caused by unwanted feedback between stages. (20-1) 2. _____
3. As long as the operation of an op amp is linear, the error voltage approaches ___. (20-1) 3. _____
4. An audio amplifier ideally amplifies signals with frequencies from 20 Hz to ___ kHz. (20-1) 4. _____
5. The closed-loop voltage gain of an IVA equals the feedback resistor divided by the ___ resistor. (20-2) 5. _____
6. The closed-loop gain-bandwidth product of an IVA is constant when A_{CL} is much greater than ___. (20-2) 6. _____
7. The closed-loop bandwidth of an IVA equals f_{unity} times B, when AB is much greater than ___. (20-2) 7. _____
8. The closed-loop impedance of an IVA equals the ___ resistor. (20-2) 8. _____
9. In a JFET-controlled switchable inverter, when the gate voltage is at $V_{GS(off)}$, the JFET switch is ___. (20-3) 9. _____
10. In a summing amplifier, all the input currents flow through the ___ resistor. (20-4) 10. _____
11. An inverting voltage amplifier with several inputs, each with a different voltage gain, that allows us to combine the signals from different sources is called a ___. (20-4) 11. _____
12. When an emitter follower is used as a current booster for an op amp, the current gain of the transistor causes the maximum load current to increase by a factor of ___. (20-5) 12. _____
13. The output current is independent of the load resistance when the load is being driven by a ___ current source. (20-6) 13. _____
14. Manufacturers can put voltage followers and differential amplifiers on a single chip to get an ___ instrumentation amplifier. (20-7) 14. _____
15. An instrumentation amplifier has a high CMRR and a ___ input impedance. (20-7) 15. _____
16. Mathematically, each pole in an active filter produces one ___ factor in the transfer function. (20-8) 16. _____
17. Another name for the Butterworth filter is a maximally ___ filter. (20-8) 17. _____
18. Above the cutoff frequency of a one-pole low-pass filter, the output decreases at a rate of ___ dB per octave. (20-8) 18. _____
19. Below the cutoff frequency of a three-pole high-pass filter, the output decreases at a rate of ___ dB per decade. (20-8) 19. _____
20. Above the cutoff frequency of a two-pole low-pass filter, the output decreases at a rate of ___ dB per decade. (20-8) 20. _____

III. ILLUSTRATIVE AND PRACTICE PROBLEMS

Illustrative Problem 1. The op amp in Fig. 20-1 has $z_{in} = 2$ MΩ, $z_{out} = 100$ Ω, and $f_{unity} = 5$ MHz. If $A = 100{,}000$, determine the following:

a. A_{CL}
b. v_{out}
c. v_{error}
d. $f_{2(CL)}$
e. $z_{in(CL)}$
f. $z_{out(CL)}$

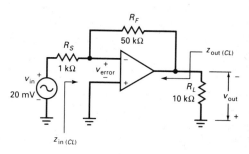

Figure 20-1

Steps	Comments
1. Find A_{CL}. $\quad A_{CL} = \dfrac{50 \text{ k}\Omega}{1 \text{ k}\Omega} = 50$	Closed-loop voltage gain. $\quad A_{CL} = \dfrac{R_F}{R_S}$
2. Find v_{out}. $\quad v_{out} = 50(20 \text{ mV}) = 1$ V	Output voltage. $\quad v_{out} = A_{CL} v_{in}$
3. Find $f_{2(CL)}$. $\quad f_{2(CL)} = \dfrac{5 \text{ MHZ}}{50} = 100{,}000$ Hz	Closed-loop bandwidth. $\quad f_{2(CL)} = f_{unity}/A_{CL}$ when $A_{CL} \gg 1$
4. Find v_{error}. $\quad v_{error} = \dfrac{1 \text{ V}}{100{,}000} = 10$ μV	Error voltage. $\quad v_{error} = \dfrac{v_{out}}{A}$
5. Find $z_{in(CL)}$. $\quad z_{in(CL)} = 1$ kΩ	Closed-loop input impedance. $\quad z_{in(CL)} = R_S$ The Miller effect acting on R_F causes the inverting terminal of the op amp to become a virtual ground.
6. Find $z_{out(CL)}$. $\quad z_{out(CL)} = \dfrac{100 \text{ }\Omega}{(1 + 1960)} = 0.051$ Ω	Closed-loop output impedance. $\quad z_{out(CL)} = \dfrac{z_{out}}{1 + AB}$ $B = \dfrac{R_S}{R_F + R_S} = \dfrac{1 \text{ k}\Omega}{51 \text{ k}\Omega} = 0.0196$ $AB = 100{,}000 (0.0196) = 1960$

Practice Problem 1. The op amp in Fig. 20-2 has $z_{in} = 1$ MΩ, $z_{out} = 200$ Ω, and $f_{unity} = 4$ MHz. If $A = 200{,}000$, determine the following:

a. A_{CL} d. $f_{2(CL)}$
b. v_{out} e. $z_{in(CL)}$
c. v_{error} f. $z_{out(CL)}$

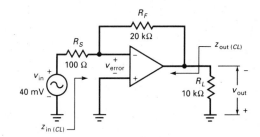

Figure 20-2

Answers: **a.** $A_{CL} = 200$ **b.** $v_{out} = 8$ V **c.** $v_{error} = 40$ μV
d. $f_{2(CL)} = 20$ kHz **e.** $z_{in(CL)} = 100$ Ω **f.** $z_{out(CL)} = 0.2$ Ω

IV. PROBLEMS

Sec. 20-1 Noninverting Voltage Amplifiers

20-1. For the circuit of Fig. 20-3, determine the following:
   a. The output voltage in the midband.
   b. The upper closed-loop critical frequency.
   c. The three lower critical frequencies.

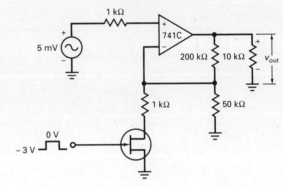

Figure 20-3

20-2. If the JFET in the circuit of Fig. 20-4 has $V_{GS(off)} = -3$ V and a drain resistance of 25 Ω when turned on, determine the following:
   a. The maximum output voltage.
   b. The minimum output voltage.

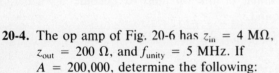

Figure 20-4

Sec. 20-2 The Inverting Voltage Amplifier

20-3. The op amp of Fig. 20-5 has $z_{in} = 2$ MΩ, $z_{out} = 100$ Ω, and $f_{unity} = 5$ MHz. If $A = 100{,}000$, determine the following:
   a. B
   b. A_{CL}
   c. v_{out}
   d. v_{error} (exact)
   e. $z_{in(CL)}$
   f. $z_{out(CL)}$
   g. $f_{2(CL)}$

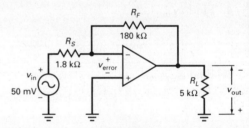

Figure 20-5

20-4. The op amp of Fig. 20-6 has $z_{in} = 4$ MΩ, $z_{out} = 200$ Ω, and $f_{unity} = 5$ MHz. If $A = 200{,}000$, determine the following:
   a. v_{out}
   b. $z_{in(CL)}$
   c. $z_{out(CL)}$
   d. The closed-loop bandwidth.

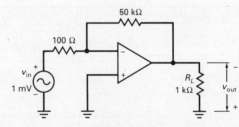

Figure 20-6

20-5. For the circuit of Fig. 20-7, do the following:
   a. Complete Table 20-1 for the given circuit conditions.

R_S kΩ	R_F kΩ	A	v_{out} V	v_{error} μV
1	50	100,000		
2	50	100,000		
1	100	100,000		
1	50	200,000		

Table 20-1

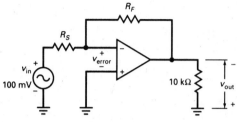

Figure 20-7

Select either *increased*, *decreased*, or *stayed the same* as the correct answer to the following questions.
   b. What happened to v_{out} when only R_S was increased?
   c. What happened to v_{out} when only R_F was increased?
   d. What happened to v_{out} when only A was increased?
   e. What happened to v_{error} when only R_S was increased?
   f. What happened to v_{error} when only R_F was increased?
   g. What happened to v_{error} when only A was increased?

20-6. What is the output voltage for the circuit of Fig. 20-8?

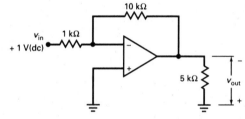

Figure 20-8

20-7. For the circuit of Fig. 20-9, if A = 200,000, determine the following:
   a. v_{out}
   b. The current through the 100-Ω resistor.
   c. The current through the 5-kΩ resistor.

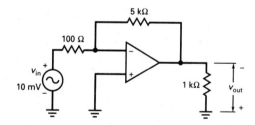

Figure 20-9

20-8. The circuit of Fig. 20-10 has A = 300,000. If $z_{in(CL)}$ = 1 kΩ, find the following:
   a. R_S
   b. R_F

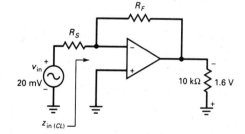

Figure 20-10

Sec. 20-3 Op-Amp Inverting Circuits

20-9. For the circuit of Fig. 20-11, determine the following:
  a. $A_{(CL)}$
  b. v_{out}
  c. $f_{2(CL)}$

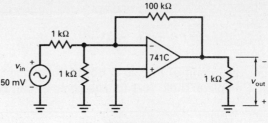

Figure 20-11

20-10. For the circuit of Fig. 20-12, find the following:
  a. The critical frequency of C_{in}.
  b. The critical frequency of C_{out}.
  c. The critical frequency of C_{BY}.
  d. v_{out} at midband

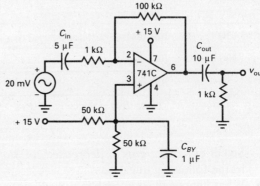

Figure 20-12

Sec. 20-4 The Summing Amplifiers

20-11. What is the output voltage in the circuit of Fig. 20-13?

20-12. What is the output voltage in the circuit of Fig. 20-14?

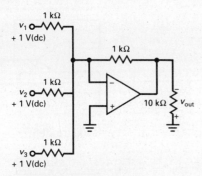

Figure 20-13

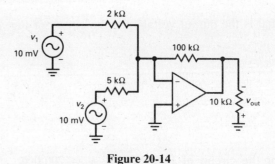

Figure 20-14

Sec. 20-5 Current Boosters for the Voltage Amplifiers

20-13. In the circuit of Fig. 20-15, $\beta_{dc} = 200$ for the transistor. If $R_L = 100\ \Omega$, determine the following:
  a. v_{out}
  b. The load current.
  c. The base current.

20-14. In the circuit of Fig. 20-15, $\beta_{dc} = 200$ for the transistor. If $R_L = 10\ \Omega$, determine the following:
  a. v_{out}
  b. The load current.
  c. The base current.

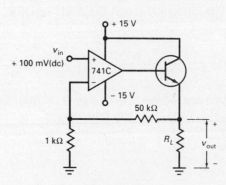

Figure 20-15

Sec. 20-6 Voltage Controlled Current Sources

20-15. For the circuit of Fig. 20-16, do the following:
  a. Complete Table 20-2 for the given circuit conditions.

R kΩ	V_{in}(dc) V	R_L Ω	i_{out} mA(dc)
1	1	100	
2	1	100	
1	2	100	
1	1	200	

Table 20-2

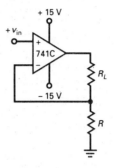

Figure 20-16

Select either *increased*, *decreased*, or *stayed the same* as the correct answer to the following questions.
  b. What happened to i_{out} when only R was increased?
  c. What happened to i_{out} when only v_{in} was increased?
  d. What happened to i_{out} when only R_L was increased?

20-16. For the circuit of Fig. 20-17, do the following:
  a. Complete Table 20-3 for the given circuit conditions.

R kΩ	V_{in}(dc) V	R_L Ω	i_{out} mA(dc)
1	5	100	
2	5	100	
1	10	100	
1	5	200	

Table 20-3

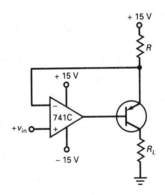

Figure 20-17

Select either *increased*, *decreased*, or *stayed the same* as the correct answer to the following questions.
  b. What happened to i_{out} when only R was increased?
  c. What happened to i_{out} when only v_{in} was increased?
  d. What happened to i_{out} when only R_L was increased?

CHAPTER 21

NONLINEAR OP-AMP CIRCUITS

Study Chap. 21 in *Electronic Principles*.

I. TRUE / FALSE

Answer true (T) or false (F) to each of the following statements.

Answer

1. The shape of the output signal in a nonlinear op-amp circuit is different from the input. **(21-1)** 1. ___
2. The op amp never saturates in a nonlinear op-amp circuit. **(21-1)** 2. ___
3. An active half-wave rectifier can rectify small signals whose peak values are much less than 0.7 V. **(21-1)** 3. ___
4. The discharge time constant of an active peak detector is much longer than the period of the signal. **(21-1)** 4. ___
5. The output of an active positive limiter has the same size and shape as the input signal except for a positive dc shift. **(21-1)** 5. ___
6. A comparator is always used as a nonlinear circuit. **(21-2)** 6. ___
7. If the slew rate of an op amp is increased, the time it needs to switch between a high to a low output will increase. **(21-2)** 7. ___
8. A flag is a signal which indicates that an event has taken place. **(21-2)** 8. ___
9. A comparator produces a two-state output. **(21-2)** 9. ___
10. A way to increase the switching time of an op amp that is used as a comparator is to eliminate the compensating capacitor. **(21-2)** 10. ___
11. When a large sine wave drives a Schmitt trigger, the output is a triangular wave. **(21-3)** 11. ___
12. If the peak-to-peak noise voltage in a Schmitt trigger is less than the hysteresis voltage, then the noise cannot produce false triggering. **(21-3)** 12. ___
13. The hysteresis voltage of a Schmitt trigger is equal to the difference of its upper and lower trip points. **(21-3)** 13. ___
14. For an integrator to work properly, the closed-loop time constant of the input bypass circuit should be at least 10 times greater than the width of the input pulse. **(21-4)** 14. ___
15. The closed-loop time constant of a typical op-amp integrator is extremely long due to the Miller effect. **(21-4)** 15. ___
16. If the input waveform of an integrator is rectangular, then its output will be sinusoidal. **(21-4)** 16. ___

17. Op amps can be used to convert sine waves to rectangular waves, rectangular waves to triangular waves, and triangular waves to pulses. (21-5)

17. ___

18. The frequency of the output voltage in a Schmitt trigger is greater than its input frequency. (21-5)

18. ___

19. If the input to an integrator is a rectangular wave with an average value of zero, then the average value of the output is also zero. (21-5)

19. ___

20. If the RC time constant of a relaxation oscillator is increased, then its frequency will decrease. (21-6)

20. ___

II. COMPLETION

Complete each of the following statements.

Answer

1. The high gain of an op amp in an active diode circuit almost eliminates the effect of the ____ voltage on the diode. (21-1)

2. The discharge time constant of an active peak detector should be at least ____ times longer than the period of the lowest input frequency. (21-1)

3. The discharge time constant of an active peak detector with a small load can be increased by using an op amp as a ____. (21-1)

4. If the reference voltage is zero in an active positive limiter, then the output will be equal to the ____ portion of the input signal. (21-1)

5. The output of an active positive clamper has the same size and shape as the input signal except for a ____ dc shift. (21-1)

6. The input voltage of a comparator that causes the output to switch states (low to high, or vice versa) is called the ____ point. (21-2)

7. A comparator with a reference voltage of zero is often called a ____-____ detector. (21-2)

8. A comparator whose positive output indicates that the input voltage exceeds a specific limit is sometimes called a ____ detector. (21-2)

9. The resistor in the output portion of the comparator that causes the output to be pulled up to the supply voltage is called a ____ resistor. (21-2)

10. The voltage level of the output drive in a comparator ideal for TTL devices is either 0 or +____ V. (21-2)

11. A Schmitt trigger is a comparator used with ____ feedback. (21-3)

12. The number of trip points a Schmitt trigger has is ____. (21-3)

13. A useful property that a Schmitt trigger utilizes to prevent false triggering associated with unwanted noise voltages is called ____. (21-3)

14. A circuit that performs a mathematical operation called integration is called an ____. (21-4)

15. A voltage that is linearly increasing or decreasing is called a ____. (21-4)

16. In an op-amp integrator the feedback component is a ____. (21-4)

17. If the frequency of an integrator's rectangular input voltage is increased, then the peak-to-peak value of its output voltage will ____. (21-5)

18. The duty cycle is defined as the width of the pulse divided by the ____. (21-5)

19. If the feedback capacitor of an integrator with a rectangular input voltage is decreased, then the peak-to-peak value of its output voltage will ____. (21-5)

20. A circuit that generates an output signal whose frequency depends on a charging capacitor is called a ____ oscillator. (21-6)

CHAPTER 22
OSCILLATORS

Study Chap. 22 in *Electronic Principles*.

I. TRUE / FALSE

Answer true (T) or false (F) to each of the following statements.

Answer

1. After the desired output is reached in any feedback oscillator, the loop gain AB must decrease to zero. **(22-1)** 1. ___
2. Above and below the resonant frequency of an oscillator, the phase shift around the feedback loop is either above or below 0°. **(22-1)** 2. ___
3. The voltage that starts an oscillator is caused by the noise voltage in the resistors. **(22-1)** 3. ___
4. A Wein-bridge oscillator uses only positive feedback. **(22-2)** 4. ___
5. Coupling and bypass circuits are examples of phase-shifting circuits. **(22-2)** 5. ___
6. Sinusoidal oscillators always use some kind of phase-shifting circuit to produce oscillation at one frequency. **(22-2)** 6. ___
7. If the capacitors of a Wein-bridge oscillator are increased, then the resonant frequency will increase. **(22-2)** 7. ___
8. A twin-T oscillator cannot be easily adjusted to operate over a large frequency range as a Wein-bridge oscillator can. **(22-3)** 8. ___
9. A phase-shift oscillator usually has either two lead or lag circuits in its feedback path. **(22-3)** 9. ___
10. The feedback voltage necessary for oscillation in a Colpitts oscillator is produced by an inductive voltage divider. **(22-4)** 10. ___
11. Op amps are not used as the amplifier in a Colpitts oscillator because the resonant frequency of the LC oscillator is beyond the f_{unity} of most op amps. **(22-4)** 11. ___
12. Capacitive or link coupling is used in a Colpitts oscillator to prevent excessive loading of the tank circuit. **(22-4)** 12. ___
13. The output signal of a Hartley oscillator can either be capacitively or link coupled. **(22-5)** 13. ___
14. The transistor and stray capacitance in a Clapp oscillator have a great effect on the oscillating frequency. **(22-5)** 14. ___
15. If the thickness of the crystal is increased, its resonant frequency would increase. **(22-6)** 15. ___
16. Crystals have a very high Q. **(22-6)** 16. ___

17. The oscillation frequency of a crystal will be between its series and parallel resonant frequency. (22-6)   17. ____

18. The crystal mainly acts as a capacitor in a Colpitts crystal oscillator. (22-6)   18. ____

19. The best solution for an amplifier with a motorboating problem is to use a power supply with an extremely small Thevenin resistance. (22-7)   19. ____

20. Unwanted capacitive and magnetic feedback can be controlled by properly shielding each stage of the amplifier. (22-7)   20. ____

II. COMPLETION

Complete each of the following statements.

Answer

1. In order to get maximum feedback at one frequency, the feedback circuit of an oscillator is usually a ____ circuit. (22-1)   1. _____

2. The initial loop gain of any feedback oscillator must be greater than ____ at the frequency at which the loop phase shift is 0°. (22-1)   2. _____

3. The initial voltage that causes an oscillator to begin to oscillate is called the ____ voltage. (22-1)   3. _____

4. The feedback fraction of the resonant circuit in a Wein-bridge oscillator is equal to ____. (22-2)   4. _____

5. The Wein-bridge oscillator uses a feedback circuit called a ____-____ circuit to produce oscillations at one frequency. (22-2)   5. _____

6. The phase angle of the resonant circuit in a Wein-bridge oscillator is equal to ____ degrees. (22-2)   6. _____

7. The closed-loop voltage gain from the noninverting input to the output of the Wein-bridge oscillator is greater than ____ when the power is first turned on. (22-2)   7. _____

8. The twin-T filter is sometimes referred to as a ____ filter. (22-3)   8. _____

9. If the resistance of a twin-T filter is increased, the resonant frequency will ____. (22-3)   9. _____

10. An *LC* oscillator is a circuit that can be used for frequencies between 1 and ____ MHz. (22-4)   10. _____

11. Most *LC* oscillators use tank circuits with a *Q* greater than ____. (22-4)   11. _____

12. One of the most widely used *LC* oscillators is a ____ oscillator. (22-4)   12. _____

13. An Armstrong oscillator uses ____ coupling for the feedback signal. (22-5)   13. _____

14. The feedback voltage necessary for oscillations in a Hartley oscillator is produced by an ____ voltage divider. (22-5)   14. _____

15. A phenomenon when a material is mechanically forced to vibrate and it generates an ac voltage at the same frequency is called the ____ effect. (22-6)   15. _____

16. The lowest resonant frequency of a crystal is called the ____ frequency. (22-6)   16. _____

17. Because it is inexpensive and readily available in nature, the crystal material that is widely used for *RF* oscillators and filters is ____. (22-6)   17. _____

18. The higher resonant frequencies of a crystal are called ____. (22-6)   18. _____

19. Very high unwanted oscillations that are caused by the small transistor capacitance and lead inductances distributed throughout the circuit are called ____ oscillations. (22-7)   19. _____

20. An unwanted ac potential difference between two ground points that can cause high-frequency oscillations in an amplifier is called a ____ loop. (22-7)   20. _____

CHAPTER 23
REGULATED POWER SUPPLIES

Study Chap. 23 in *Electronic Principles*.

I. TRUE / FALSE

Answer true (T) or false (F) to each of the following statements.

Answer

1. A zener diode with a positive temperature coefficient provides a constant zener voltage over a large range of temperature changes. (23-1)   1. ___
2. A Darlington pair can be used for the pass transistor in a series regulator so that the regulator can drive a smaller load resistor. (23-1)   2. ___
3. Zener diodes with breakdown voltages between 5 and 6 V have temperature coefficients of approximately zero. (23-1)   3. ___
4. When the load terminals are shorted in a series regulator, simple current limiting does not protect the pass transistor from dissipating a large amount of power. (23-2)   4. ___
5. A voltage regulator with simple current limiting can handle much larger lead currents than one with foldback current limiting. (23-2)   5. ___
6. The current-sensing resistor limits the load current in a voltage regulator. (23-2)   6. ___
7. The output impedance of a regulated power supply is very small. (23-3)   7. ___
8. Ripple rejection is defined as the input ripple voltage divided by the output ripple voltage. (23-3)   8. ___
9. A voltage regulator attenuates the ripple voltage that comes in with the unregulated input voltage. (23-3)   9. ___
10. When an IC voltage regulator is more than a few inches from the filter capacitor of an unregulated input, the lead inductance may produce oscillations within the IC. (23-4)   10. ___
11. The LM320 series is a group of positive voltage regulators. (23-4)   11. ___
12. Any device in the LM 340 series needs a minimum input voltage at least 2 to 3 V greater than the regulated output voltage. (23-4)   12. ___
13. The maximum input voltage of an IC voltage regulator should not be exceeded because of excessive power dissipation. (23-4)   13. ___
14. A current booster is a transistor in series with the IC regulator. (23-5)   14. ___
15. A current-limiting transistor can be used to limit the current through an outboard transistor. (23-5)   15. ___

16. A phase splitter produces two output voltages that are equal in amplitude and are opposite in phase. (23-6)   16. ___
17. The output voltage of a dc-to-dc converter can only be greater than the input voltage. (23-6)   17. ___
18. Because the pass transistor in a switching regulator operates only in the saturation and cutoff region, it dissipates less power. (23-7)   18. ___
19. In a stepdown switching regulator, the output voltage is directly proportional to the duty cycle of the pulses that drives the pass transistor. (23-7)   19. ___
20. The widths of the pulses that drive the pass transistor in the stepdown switching regulator are controlled by negative feedback. (23-7)   20. ___

II. COMPLETION

Complete each of the following statements.

Answer

1. The zener voltage in a voltage regulator is called the ___ voltage. (23-1)
2. Voltage regulation can be improved by using ___ feedback. (23-1)
3. In a series regulator, all the load current flows through the ___ transistor. (23-1)
4. A voltage regulator without some form of current limiting will have no ___-___ protection. (23-2)
5. A dominant bypass circuit in a voltage regulator will prevent ___. (23-2)
6. The short-circuit load current is much smaller than the maximum load current in a voltage regulator with ___ current limiting. (23-2)
7. Load regulation is also called ___ effect. (23-3)
8. V_{NL} occurs when the load is ___. (23-3)
9. Source regulation is also called ___ regulation. (23-3)
10. The LM 340-5 produces a regulated output voltage of + ___ V. (23-4)
11. The term used to indicate that an IC voltage regulator will automatically turn itself off if the internal temperature becomes dangerously high is called ___ shutdown. (23-4)
12. The term used to indicate that an IC voltage regulator has stopped regulating because the input voltage is less than the minimum is called ___. (23-4)
13. The minimum allowable difference between the input voltage and the output voltage of an IC voltage regulator is called the ___-___ voltage. (23-4)
14. The transistor put across an IC regulator to increase the load current capability is called an ___ transistor. (23-5)
15. If the current-sensing resistor of a current-limiting transistor is increased, then the limiting current will ___. (23-5)
16. In most dc-to-dc converters, the input dc voltage is applied to a square-wave oscillator whose output drives a ___. (23-6)
17. The frequency that works best with a dc-to-dc converter that uses a square-wave oscillator is ___ kHz. (23-6)
18. Because series regulators have their pass transistor operating in the active region, they are sometimes called ___ regulators. (23-7)
19. The diode used in the stepdown switching regulator protects the pass transistor from the kickback voltage of the ___. (23-7)
20. The three basic types of switching regulators are the stepdown, step-up, and ___. (23-7)

CHAPTER 24

COMMUNICATIONS CIRCUITS

Study Chap. 24 in *Electronic Principles*.

I. TRUE / FALSE

Answer true (T) or false (F) to each of the following statements.

Answer

1. To avoid distortion, a class C amplifier always drives a resonant tank circuit. (24-1)   1. ___
2. A resonant tank circuit with a Q greater than 10 has a very low impedance. (24-1)   2. ___
3. When a load is put across a resonant tank circuit, the overall Q of the circuit will increase. (24-1)   3. ___
4. A high-Q resonant tank circuit can be used to restore the fundamental frequency of a nonsinusoidal waveform. (24-1)   4. ___
5. The dc value of the collector current in a class C amplifier does not depend on the conduction angle of the transistor. (24-2)   5. ___
6. Class C amplifiers obtain maximum load power and maximum efficiency when they use the entire ac load line. (24-2)   6. ___
7. The efficiency of a frequency multiplier increases when the class C circuit is tuned to higher harmonics. (24-3)   7. ___
8. The Q of the resonant circuit in a tuned class C circuit has no effect on the shape of the load voltage. (24-3)   8. ___
9. The dc component is the average value of the periodic wave. (24-4)   9. ___
10. Any periodic nonsinusoidal waveform is equivalent to a dc component, a fundamental frequency, and harmonics. (24-4)   10. ___
11. The frequency mixer in an AM receiver is a linear device. (24-5)   11. ___
12. The frequency mixer in an AM receiver includes a filter that blocks all frequencies except the difference frequency. (24-5)   12. ___
13. The noise voltage of a resistor will increase if the temperature of the resistor increases. (24-6)   13. ___
14. The noise voltage of a resistor cannot be calculated. (24-6)   14. ___
15. The upper and lower envelopes of an AM waveform have the shape of the modulating signal. (24-7)   15. ___
16. Percent modulation is used to measure the amount of amplitude modulation. (24-8)   16. ___
17. An AM signal produced by a carrier and a modulating signal has only two frequency components. (24-9)   17. ___

18. In a superheterodyne receiver, audio amplifiers amplify the modulating signal. (24-11)

18. ____

19. In an FM modulator, the frequency deviation is directly proportional to the modulating signal. (24-12)

19. ____

20. The major advantage of AM over FM is that the reception is less noisy. (24-12)

20. ____

II. COMPLETION

Complete each of the following.

Answer

1. Class C operations mean that the collector current flows for less than ____ degrees of the ac cycle. (24-1)

1. _____

2. The bandwidth of a resonant tank circuit equals the resonant frequency divided by the ____ factor of the entire circuit. (24-1)

2. _____

3. Class C amplifiers with bandwidths that are less than 10 percent of the resonant frequency are called ____ amplifiers. (24-1)

3. _____

4. The maximum efficiency of a class C amplifier is ____ percent. (24-2)

4. _____

5. When the conduction angle of the transistor in a class C amplifier is decreased, the stage's efficiency will ____. (24-2)

5. _____

6. If the frequency of the input signal to a class C multiplier is 1 MHz, the fundamental frequency of the collector-currents narrow pulses is ____ Hz. (24-3)

6. _____

7. Nonlinear distortion of a signal with respect to the frequency domain is called ____ distortion. (24-4)

7. _____

8. The fundamental frequency is sometimes called the ____ harmonic. (24-4)

8. _____

9. The frequency-domain instrument that displays the harmonic peak values versus frequency of a periodic voltage waveform is called a ____ analyzer. (24-4)

9. _____

10. The output frequency of the mixer in an AM receiver is ____ kHz. (24-5)

10. _____

11. In an AM receiver, one of the input signals to the mixer comes from a circuit called a ____ oscillator. (24-5)

11. _____

12. The kind of noise that is produced inside a resistor is called ____ noise. (24-6)

12. _____

13. Noise produced by the vibrations that cause capacitor plates or inductor windings to move is called ____. (24-6)

13. _____

14. The high-frequency input signal to an AM modulator is called the ____. (24-7)

14. _____

15. The low-frequency input signal to an AM modulator is called the ____ signal. (24-7)

15. _____

16. The modulation coefficient of an AM waveform is always in the range of zero to ____. (24-8)

16. _____

17. The difference component of an AM signal is sometimes called the ____ side frequency. (24-9)

17. _____

18. The circuit in an AM receiver that separates the modulating signal from the carrier is called a ____. (24-10)

18. _____

19. In a superheterodyne receiver, the frequency that the local oscillator operates at is ____ kHz above the incoming RF signal. (24-11)

19. _____

20. In an FM modulator, the difference between the oscillator frequency and the quiescent frequency is called the frequency ____. (24-12)

20. _____

ANSWERS

ANSWERS TO TRUE/FALSE QUESTIONS

Chapter 1
1. T
2. T
3. T
4. F
5. T
6. F
7. T
8. F
9. T
10. T
11. T
12. F
13. T
14. F
15. T
16. T
17. T
18. F
19. T

Chapter 2
1. T
2. F
3. T
4. F
5. F
6. T
7. T
8. T
9. F
10. T
11. F
12. F
13. F
14. T
15. F
16. T
17. T
18. F
19. F
20. F
21. T
22. T
23. F
24. F
25. T
26. T
27. T
28. F
29. T
30. T
31. T
32. F
33. T
34. T
35. T
36. T
37. T
38. F
39. T
40. T

Chapter 3
1. F
2. T
3. T
4. T
5. T
6. T
7. T
8. T
9. T
10. T
11. F
12. T
13. T
14. F
15. F
16. F
17. T
18. T
19. T
20. F
21. F

Chapter 4
1. F
2. T
3. F
4. F
5. F
6. F
7. T
8. T
9. T
10. T
11. T
12. T
13. F
14. T
15. T
16. T
17. T
18. T
19. T
20. T

Chapter 5
1. F
2. T
3. T
4. T
5. T
6. F
7. T
8. F
9. T
10. F
11. T
12. F
13. T
14. T
15. T
16. T
17. F
18. T
19. T
20. T

Chapter 6
1. F
2. T
3. T
4. T
5. F
6. T
7. F
8. T
9. F
10. T
11. T
12. F
13. T
14. T
15. T
16. T
17. T
18. T
19. T
20. T
21. T
22. T
23. T
24. F

Chapter 7
1. F
2. T
3. T
4. T
5. F
6. T
7. T
8. T
9. T
10. F
11. T
12. T
13. F
14. F
15. T
16. T
17. T
18. T
19. T
20. F

Chapter 8
1. T
2. T
3. T
4. T
5. F
6. T
7. T
8. T
9. T
10. F

11. F
12. T
13. F
14. F
15. T
16. F
17. F
18. F
19. T
20. T

Chapter 9
1. F
2. T
3. T
4. F
5. T
6. T
7. T
8. T
9. T
10. T
11. T
12. T
13. T
14. T
15. T
16. T
17. T
18. F
19. T

Chapter 10
1. T
2. F
3. T
4. T
5. T
6. F
7. T
8. F
9. F
10. T
11. T
12. T
13. T
14. F
15. T
16. F
17. T
18. F
19. T
20. T

Chapter 11
1. T
2. F
3. T
4. F
5. F
6. T
7. T
8. F

9. T
10. F
11. T
12. T
13. F
14. T
15. T
16. T
17. T
18. F
19. T
20. T

Chapter 12
1. F
2. T
3. T
4. F
5. F
6. F
7. T
8. T
9. T
10. F
11. T
12. T
13. F
14. F
15. T
16. T
17. T
18. T
19. F
20. T
21. T

Chapter 13
1. T
2. T
3. F
4. F
5. T
6. T
7. F
8. T
9. T
10. T
11. T
12. T
13. T
14. T
15. F
16. T
17. T
18. F
19. T
20. T

Chapter 14
1. T
2. T

3. T
4. F
5. F
6. T
7. T
8. F
9. T
10. T
11. T
12. T
13. T
14. T
15. T
16. T
17. T
18. T
19. F
20. T

Chapter 15
1. F
2. T
3. T
4. T
5. F
6. T
7. T
8. F
9. T
10. T
11. T
12. T
13. T
14. F
15. F
16. T
17. T
18. F
19. T
20. T

Chapter 16
1. T
2. T
3. F
4. T
5. F
6. T
7. F
8. T
9. F
10. T
11. T
12. F
13. T
14. F
15. T
16. F
17. T
18. F
19. T

Chapter 17
1. F
2. F
3. T
4. T
5. T
6. T
7. T
8. F
9. F
10. T
11. T
12. F
13. T
14. F
15. T
16. T
17. T
18. F
19. F
20. T

Chapter 18
1. T
2. T
3. T
4. F
5. T
6. T
7. T
8. F
9. T
10. T
11. F
12. F
13. T
14. T
15. T
16. T
17. T
18. F
19. T
20. T

Chapter 19
1. T
2. T
3. F
4. F
5. T
6. T
7. F
8. T
9. T
10. T
11. T
12. T
13. T
14. T
15. T
16. F
17. F

Copyright © 1993 by the Glencoe Division of Macmillan/McGraw-Hill School Publishing Company. All rights reserved.

18	F	13	T	7	F	**Chapter 22**		16	T	11	F	6	T
19	T	14	T	8	T	1	F	17	T	12	F	7	F
Chapter 20		15	T	9	T	2	T	18	F	13	T	8	F
		16	F	10	T	3	F	19	T	14	F	9	T
1	T	17	F	11	F	4	F	20	T	15	T	10	T
2	T	18	T	12	T	5	T	**Chapter 23**		16	T	11	F
3	F	19	F	13	T	6	T	1	F	17	F	12	F
4	T	20	T	14	T	7	F	2	T	18	T	13	T
5	F			15	T	8	T	3	T	19	T	14	F
6	T	**Chapter 21**		16	F	9	F	4	T	20	T	15	T
7	T	1	T	17	T	10	F	5	F	**Chapter 24**		16	T
8	F	2	F	18	F	11	T	6	T	1	T	17	F
9	T	3	T	19	T	12	T	7	T	2	F	18	T
10	T	4	T	20	T	13	T	8	T	3	F	19	T
11	T	5	F			14	F	9	T	4	T	20	F
12	F	6	T			15	F	10	T	5	F		

ANSWERS TO COMPLETION QUESTIONS

Chapter 1
1. Constant
2. Stiff
3. Internal
4. Theoretical
5. Infinite
6. Ideal
7. Large
8. Open-circuit
9. Short
10. Open
11. Shorted-load
12. Thevenin
13. Thevenin
14. Infinite
15. Zero
16. Opened
17. Shorted
18. Ideal or first
19. Third
20. Zero
21. Second

Chapter 2
1. Weak
2. Valance
3. Ion
4. Four
5. Valance
6. Eight
7. Covalent
8. Recombination
9. Lifetime
10. Thermal
11. Intrinsic
12. Valance
13. Carriers
14. Doping
15. Acceptor
16. Donor
17. n
18. p
19. Majority
20. Majority
21. Minority
22. Junction
23. Depletion
24. Silicon
25. 0.3
26. 0.7
27. Increases
28. Zero
29. Saturation
30. Surface-leakage
31. Avalanche
32. Zener
33. 2 mV
34. 10

Chapter 3
1. Anode
2. Cathode
3. Forward
4. Nonlinear
5. Data
6. Bulk
7. Knee
8. Breakdown
9. Open
10. 0.7 V
11. Conduct
12. Bulk
13. Small
14. First
15. Shorted
16. Opened
17. Dependent
18. Breakdown
19. Maximum

Chapter 4
1. Up
2. Up
3. 31.8
4. dc
5. ½
6. Grounded
7. 4
8. 2
9. Capacitor
10. Decrease
11. 10
12. Decrease
13. Secondary
14. 50
15. Surge
16. Ripple
17. Average
18. Oscilloscope
19. V_{RRM}
20. I_{SFM}

Chapter 5
1. Regulator
2. Battery
3. Regulators
4. Series
5. Coefficient
6. Zero
7. Photodiode
8. Light
9. Optocoupler
10. Schottley
11. Recovery
12. Charge
13. Tuning
14. Capacitor
15. Transient
16. Zener
17. Derating
18. Impedance
19. Tolerance
20. Voltages

Chapter 6
1. 2
2. Base
3. n
4. n
5. Diodes
6. Reverse
7. Base
8. Base
9. Emitter
10. Emitter
11. Base
12. V_{CE}
13. Base
14. Diode
15. Cutoff
16. Saturation
17. Active
18. Collector
19. Active
20. Bulk
21. Saturation
22. Opened
23. Heat
24. h

Chapter 7
1. h_{FE}
2. Tolerances
3. $I_{C(sat)}$
4. Shorted
5. $V_{CE(cut)}$
6. Opened
7. Quiescent
8. Q
9. Experimentally
10. Contradiction
11. Hard
12. Soft
13. Switching
14. Emitter
15. Base
16. Active
17. Hard
18. Independent
19. Dependent
20. 1000

Chapter 8
1. Emitter
2. One
3. Emitter
4. 20
5. Stiff
6. Decrease
7. Decrease
8. Increase
9. Increase
10. Reverse
11. Emitter
12. 0
13. Decrease
14. −0.7
15. Negative
16. Negative
17. Less
18. Less
19. Feedback
20. Voltage-divider

Chapter 9
1. Decrease
2. 10
3. Critical
4. Short
5. Zero
6. ac
7. ac
8. Open
9. Short
10. 10
11. 25
12. Decrease
13. β
14. ac
15. 180
16. Bypass
17. Base
18. Emitter
19. h_{fe}
20. r'

Chapter 10
1. Base
2. Emitter
3. β
4. ac
5. ac
6. Amplification
7. Output
8. r'_e
9. Decrease
10. Predicted
11. Feedback
12. Stabilize
13. Swamps
14. R_L
15. r'_e
16. Cascading
17. Product
18. Increase
19. Decrease
20. Decrease

Chapter 11
1. Intercept
2. dc
3. ac
4. Steeper
5. MPP
6. Output
7. V_{CEQ}
8. 2 V_{CEQ}
9. ac
10. 360
11. Output
12. 8
13. I_{CQ}
14. Gain
15. P_S
16. 25
17. Drain
18. η
19. Ambient
20. Heat

Chapter 12
1. Common-collector
2. Output
3. Negative
4. R_L
5. r'_e
6. One
7. Low
8. Low
9. Load
10. V_{CEQ}
11. $I_{CQ}r_e$
12. r_e
13. Clipped
14. Direct
15. Capacitively
16. Darlington
17. 1.4
18. 180
19. Runaway
20. Compensating

Chapter 13
1. Base
2. Collector

3 Reverse
4 Voltage
5 Drain
6 p
7 Width
8 Breakdown
9 Pinchoff
10 Ohmic
11 Zero
12 V_{GS}
13 Parabola
14 I_{DSS}
15 Proportional
16 V_P'
17 IGFET
18 Substrate
19 I_{DSS}
20 Threshold
21 Zero
22 Knee
23 $V_{GS(off)}$
24 $V_{GS(th)}$

Chapter 14
1 Source
2 R_S
3 Gate
4 Reversed
5 Two
6 Origin
7 Quiescent
8 One
9 R_S
10 R_S/R_{DS}
11 Siemen
12 Zero
13 g_m
14 One
15 V_{GS}
16 Ohmic
17 Gate-source
18 Excellent
19 Active-load
20 Decreases
21 V_{GS}

Chapter 15
1 Trigger
2 Breakover
3 Shockley
4 Saturation
5 Holding
6 0.7
7 Three
8 Blocking
9 RC
10 Inductor
11 Crowbar
12 Light-activated
13 Gate
14 Two
15 Triac
16 Two
17 Trigger
18 Intrinsic
19 Valley
20 Functional

Chapter 16
1 Midband
2 70.7
3 0.1
4 Lower
5 0.5
6 Decrease
7 Decrease
8 Emitter
9 ac
10 100
11 Stray
12 Feedback
13 Midband
14 Dominant
15 Product
16 3
17 20
18 Bode
19 400 Hz

Chapter 17
1 Mathematical
2 Discrete
3 Substrate
4 Monolithic
5 Monolithic
6 100
7 Operational
8 50
9 Tail
10 0.7
11 Noninverting
12 Inverting
13 Offset
14 Bias
15 $2r_e'$
16 $2r_e'$
17 Offset
18 Input
19 Rejection
20 Differential

Chapter 18
1 Oscillation
2 Compensating
3 One
4 20
5 Increase
6 Increase
7 Slew
8 Increase
9 Slew
10 Power
11 S_S
12 Active
13 2,000,000
14 BIFET
15 100,000
16 75
17 V_{BE}
18 2
19 90
20 Base
21 70 V/μs

Chapter 19
1 Error
2 Loop
3 Divider
4 One
5 Infinite
6 Differential
7 Decrease
8 Decrease
9 Increase
10 Decrease
11 Desensitivity
12 Sacrifice
13 Nonlinear
14 Decrease
15 Virtual
16 Virtual
17 Zero
18 Desensitivity or sacrifice
19 Increase

Chapter 20
1 Linear
2 Oscillations
3 Zero
4 20
5 Series
6 One
7 One
8 Series
9 Opened
10 Feedback
11 Mixer
12 $β_{dc}$
13 Stiff
14 IC
15 High
16 J
17 Flat
18 6
19 60
20 40

Chapter 21
1 Knee
2 10
3 Buffer
4 Negative
5 Positive
6 Trip
7 Zero-crossing
8 Limit
9 Pullup
10 5
11 Positive
12 Two
13 Hysteresis
14 Integrator
15 Ramp
16 Capacitor
17 Decrease
18 Period
19 Increase
20 Relaxation

Chapter 22
1 Resonant
2 One
3 Starting
4 ⅓
5 Lead-lag
6 Zero
7 Three
8 Notch
9 Decrease
10 500
11 10
12 Colpitts
13 Transformer
14 Inductive
15 Piezoelectric
16 Fundamental
17 Quartz
18 Overtones
19 Parasitic
20 Ground

Chapter 23
1 Reference
2 Negative
3 Pass
4 Short-circuit
5 Oscillations
6 Foldback
7 Load
8 Infinite
9 Line
10 5
11 Thermal
12 Brownout
13 Drop-out
14 Outboard
15 Decrease
16 Transformer
17 20
18 Linear
19 Inductor
20 Inverting

Chapter 24
1 180
2 Quality
3 Narrowband
4 100
5 Increase
6 1 M
7 Harmonic
8 First
9 Spectrum
10 455
11 Local
12 Thermal
13 Microphonics
14 Carrier
15 Modulating
16 One
17 Lower
18 Demodulator
19 455
20 Deviation

ANSWERS TO PROBLEMS

Chapter 1

1a. 20 V
 b. 1.82 V, 10 V, 18.2 V, 19.8 V, 19.98 V
 c. 1000 Ω and 10,000 Ω
 d. Yes
2. $R_L > 10\ \Omega$
3a. 0.435 Ω
 b. No
4a. 0.01 V
 b. 1.49 V
 c. Yes
5a. 0.19 Ω
 b. No
6a. 1 mA
 b. 0.9999 mA, 0.999 mA, 0.99 mA, 0.909 mA, 0.5 mA
 c. $R_L < 500\ \Omega$
 d. Yes
7a. 19 kΩ
 b. No
8a. 10 V
 b. 2 Ω
9a. 20 V
 b. 1.11 Ω

10a. 60 V
  b. 8 kΩ
  c.

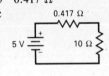

(c)

11a. 8 V
  b. 60 Ω
  c.

(c)

12a. 5 V
  b. 0.417 Ω
  c.

(c)

13a. 5 A
  b. 2 Ω
  c.

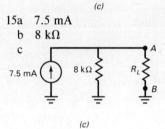

(c)

14a. 18 A
  b. 1.11 Ω
  c.

(c)

15a. 7.5 mA
  b. 8 kΩ
  c.

(c)

16a. 133 mA
  b. 60 Ω
  c.

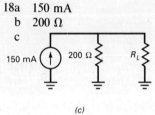

(c)

17a. 3 mA
  b. 2 kΩ
  c.

(c)

18a. 150 mA
  b. 200 Ω
  c.

(c)

19

Trouble	Voltmeter Readings		
	V_A	V_B	V_C
Circuit OK	24 V	12 V	4 V
R_1 shorted	24 V	9.6 V	0 V
R_1 opened	24 V	24 V	24 V
R_2 shorted	24 V	6 V	6 V
R_2 opened	24 V	24 V	0 V
R_3 shorted	24 V	24 V	8 V
R_3 opened	24 V	0 V	0 V
Power supply off	0 V	0 V	0 V

20

Probable Trouble	Voltmeter Readings		
	V_A	V_B	V_C
R_2 opened	0 V	30 V	30 V
R_1 shorted	15 V	30 V	30 V
Circuit OK	10 V	20 V	30 V
R_1 opened	0 V	0 V	30 V
R_3 shorted	0 V	15 V	30 V
Power supply off	0 V	0 V	0 V

Chapter 2

1.

Junction Temperature, °C	Silicon Barrier Potential, V	Germanium Barrier Potential, V
5	0.74	0.34
15	0.72	0.32
25	0.7	0.3
50	0.65	0.25
75	0.6	0.2

2

Junction Temperature, °C	Silicon Saturaton Current, nA	Germanium Saturation Current, μA
5	1	0.5
15	2	1
25	4	2
35	8	4
55	32	16

3

Reverse Voltage across the Diode, V	Silicon Surface-Leakage Current, nA	Germanium Surface-Leakage Current, nA
1	1	5
5	5	25
10	10	50
20	20	100
40	40	200

Chapter 3

1 9.3 mA
2 68 mW
3 86 Ω
4a Forward
  b Reverse
  c Reverse
  d Forward
  e Reverse
  f Forward
5a 15 mA
  b 0 V
  c 15 V
  d 0 W
  e 225 mW
  f 225 mW
6a 0 A
  b −15 V
  c 0 V
  d 0 W
  e 0 W
  f 0 W

7a 5 mA
  b 0 V
  c 5 V
  d 0 W
  e 25 mW
  f 25 mW
8a 0 A
  b −5 V
  c 0 V
  d 0 W
  e 0 W
  f 0 W
9a 2 mA
  b 0 A
  c 2 mA
  d 1.82 mA
10a 0 A
  b 2 mA
11 5 V
12 20 V
13 20 V
14 0 V

15a $V_A = 0$ V, $V_B = 0$ V
  b $V_A = 7.2$ V
      $V_B = 7.2$ V
  c $V_A = 9$ V, $V_B = 13$ V
  d $V_A = 9$ V, $V_B = 18$ V
16a 14.3 mA
  b 0.7 V
  c 14.3 V
  d 10 mW
  e 204 mW
  f 214 mW
17a 14.7 mA
  b 0.3 V
  c 14.7 V
  d 4.41 mW
  e 216 mW
  f 220 mW
18a 0 A
  b −15 V
  c 0 V
  d 0 W
  e 0 W
  f 0 W

19a 4.3 mA
  b 0.7 V
  c 4.3 V
  d 3.01 mW
  e 18.5 mW
  f 21.5 mW
20a 1.93 mA
  b 0.7 mA
  c 1.23 mA
  d 1.82 mA
21a 1.4 mA
  b 0.46 mA
22 18.6 V
23 −0.7 V
24a 42.8 mA
  b 0.721 V
  c 31.9 mW
25a 0 A
  b −5 V
  c 0 W
26 D_1

27

Probable Trouble	Voltmeter Readings	
	V_1, V	V_2, V
Diode shorted	10	0
Diode opened	10	5
Circuit OK	10	0.7 V

28

Probable Trouble	Ohmmeter Readings	
	Fig. 3-30a	Fig. 3-30b
Diode opened	∞ Ω	∞ Ω
Diode shorted	100 Ω	100 Ω
Diode OK	∞ Ω	500 Ω

29 Lead 1 is connected to the diode's cathode. Lead 2 is connected to the diode's anode.

30

	V_A	I_1	I_2	I_D
V_S increases	N	U	N	U
R_1 increases	N	D	N	D
R_2 increases	N	N	D	U

32a 2.5 V
 b 250 Ω
 c 7 mA
 d 0.75 V

31 a, b, c I_{DQ} = 7.5 mA, V_{DQ} = 0.75 V.

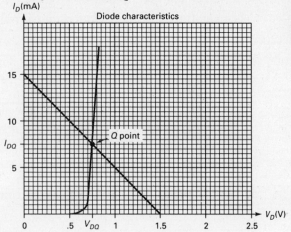

Chapter 4

1a 21.2 V
 b 120 V
 c 21.2 mA
 d 1.87 mA
 e 0.225 W
2 1:4
3a 467 V
 b 6.67 mA
 c 49.5 kΩ
 d 2.2 W
4a 3.18 V
 b 3.85 V
 c 5 V
 d 500 Hz
5a 28.3 V
 b 8.99 V
 c 2.72 mA
6a 8.48 V
 b 7.78 V

7a 14.1 V
 b 4.48 V
 c 5.43 V
 d 7.05 V
 e 2.49 mA
 f 27.6 mW
 g 121 %
 h 60 Hz
8a 63.6 V
 b 30.5 V
 c 70.7 V
 d 10 Hz
9a 21.2 V
 b 13.5 V
 c 6.14 mA
10a 8.15 V
 b 7.45 V
11a 7.08 V
 b 4.5 V
 c 2.16 V
 d 5 V
 e 25 mA

 f 139 mW
 g 48%
 h 120 Hz
12a 42.4 V
 b 27 V
 c 48.2 mA
13a 38.9 V
 b 37.5 V
14a 81.3 V
 b 51.7 V
 c 24.8 V
 d 57.4 V
 e 5.17 mA
 f 329 mW
 g 48%
 h 120 Hz
15a 170 V
 b 13.2 V
16a 2 V
 b 29 V
 c 29 V
 d 1.99%
 e 100 Hz

17a 15 mA
 b 1.25 V
 c 15.6 V
 d 0.361 V
 e 2.4%
 f 225 mW
18a 2 V
 b 0.577 V
 c 1.48%
 d 163 μF
 e 1.52 W
19 $I_{D(dc)}$ = 12.5 mA
    PIV = 14.1 V
20 $I_{D(dc)}$ = 2.59 mA
    PIV = 81.5 V

Chapter 5

1 V_{out} = 25 V, 25.45 V
   I_Z = 50 mA, 45.5 mA
2a

V_S, V	R_S, Ω	I_Z, mA	V_{out}, V
10	1 k	0	10
10	100	0	10
15	1 k	0	15
15	100	0	15
20	1 k	5	15
20	100	50	15
50	1 k	35	15
50	100	350	15

 b $V_S \geq 15$ V
 c Limits current flow through the zener diode
3 10 Ω
4a 18 V
 b 10 mA
 c 14.9 mA
 d 4.9 mA
 e 88.2 mW

5a

V_S, V	V_L, V	I_L, mA	I_S, mA	I_Z, mA
10	6	33.3	33.3	0
20	12	66.7	66.7	0
30	12	66.7	150	83.3
40	12	66.7	233	166.3
50	12	66.7	317	250.3

b $V_S \geq 12$ V

7 $R_S = 13.3\ \Omega$
8 $0.0211\ V_{pp}$
9 $11.6\ \Omega$

10a

V_S, V	V_D, V	I_D, mA
1	1	0
2	2	0
5	2	3
10	2	8
20	2	18

b Limits current flow through the LED
c $V_S > 2$ V

6a

R_L, Ω	V_L, V	I_L, mA	I_S, mA	I_Z, mA
10	0.588	58.8	58.8	0
100	4.65	46.5	46.5	0
330	10	30.3	30.3	0
470	10	21.3	30.3	9
1 k	10	10	30.3	20.3
10 k	10	1	30.3	29.3

b $R_L \geq 330\ \Omega$

11a

R_S, Ω	V_D, V	I_D, mA
150	2	120
1 k	2	18
1.2 k	2	15
1.8 k	2	10
10 k	2	1.8

b $1\ k\Omega \leq R_S \leq 1.8\ k\Omega$
12 $1\ k\Omega$
13 Yes
14 Yes
15 $17\ \Omega$

16

	Voltmeter Readings	
Trouble	V_1, V	V_2, V
Circuit OK	15	5
Zener opened	15	6
Zener shorted	15	0
R_L opened	15	5
R_L shorted	15	0
Zener inserted with polarity reversed	15	0.7
Power supply off	0	0

Chapter 6

1. 2.95 mA
2. 250
3. $I_E = 6.06$ mA
   $I_B = 60$ μA
4. a.

V_{BB}, V	R_B, Ω	I_B, μA
0.5	100 k	0
0.5	1 M	0
5	100 k	43
5	1 M	4.3
10	100 k	93
10	1 M	9.3

   b. $V_{BB} > 0.7$ V
   c. I_B increases
   d. I_B decreases

5. $R_B = 930$ kΩ
6. $R_B = 1.93$ MΩ
7. $I_B = 29.3$ μA
8. a. 53 μA
   b. 5.3 mA
   c. 5.353 mA
   d. 0.7 V
   e. 0 V
   f. 0.7 V
   g. 12.3 V
   h. 12.3 V
   i. 65.2 mW
9. a. 3.86 mA
   b. 11.5 V
   c. 44.4 mW
10. 224

11. a.

R_C, kΩ	$β_{dc}$	I_B, μA	I_C, mA	V_{CE}, V
1	100	30	3	6
1	150	30	4.5	4.5
1.8	100	30	3	3.6

   b. Remained the same
   c. Remained the same
   d. Increased
   e. Remained the same
   f. Decreased
   g. Decreased

12. A = saturation region
    B = active region
    C = breakdown region
    D = cutoff region

13. a.
    b.

R_B, Ω	R_C, kΩ	$β_{dc}$	I_B, μA	I_C, mA	V_{CE}, V	Region of Operation
Opened	1	100	0	0	15	Cutoff
220 k	1	100	65	6.5	8.5	Active
220 k	2.2	100	65	6.5	0.7	Saturation
220 k	1	225	65	14.6	0.4	Saturation
270 k	1	100	53	5.3	9.7	Active

14. a. Ideal = 3.79 mA
       Second approx. = 3.52 mA
    b. Ideal = 9.77 V
       Second approx. = 10.5 V
    c. Ideal = 37 mW
       Second approx. = 37 mW
15. a. Ideal = 6.86 mA
       Second approx. = 6.67 mA
    b. Ideal = 12.7 V
       Second approx. = 13 V
    c. Ideal = 87.1 mW
       Second approx. = 86.7 mW
16. No
17. No
18. Yes

19.

Trouble	Voltmeter Readings		
	V_1, V	V_2, V	V_3, V
Circuit OK	15	0.7	7.2
Power supply off	0	0	0
No base current	15	15	15
R_B opened	15	0	15
R_C shorted	15	0.7	15
R_C opened	15	0.7	0

Chapter 7

1a 100
 b 100 μA
 c 193 kΩ
2 270 kΩ
3 20 mA
   24 V

4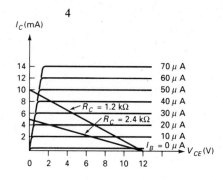

5a Remains the same
 b Remains the same
 c Remains the same
 d Decreases
 e Increases
 f Increases

6a R_C
 b V_{CC}
7 $V_{CC} = 9$ V
   $R_C = 1.5$ kΩ

8a

R_B, kΩ	R_C, kΩ	β_{dc}	I_B, μA	I_C, mA	$I_{C(sat)}$, mA	V_{CE}, V
220	1	100	51.4	5.14	12	6.86
330	1	100	34.2	3.42	12	8.58
220	1	150	51.4	7.71	12	4.29
220	1.8	100	51.4	5.14	6.67	2.75

 b Down
 c Up

9a

	Q-Point Coordinates			Circuit Values		Analytically
Q Points	I_B, μA	I_C, mA	V_{CE}, V	R_C, kΩ	R_B, kΩ	$V_C = V_{CC} - I_C R_C$, V
Q_1	25	5	1.5	1.5	332	1.5
Q_2	20	4	3	1.5	415	3
Q_3	10	2	6	1.5	830	6

 b R_B
 c Yes

10

β_{dc}	R_B, kΩ	R_C, kΩ	Voltmeter Readings		Region of Operation
			V_1, V	V_2, V	
100	180	1	0.7	7.06	Active
100	47	1	0.7	≈0	Saturation
100	180	10	0.7	≈0	Saturation
400	180	1	0.7	≈0	Saturation

11 Increase R_B
    Decrease R_C
    Decrease β_{dc}
12 Opened = 18 V
    Closed = ≈ 0 V
13a 6 V
 b 4 mA
 c 4 mA
 d 15.2 V
 e 9.2 V
 f 36.8 mW
14a 4 V
 b 12 V

15a

V_{CC}, V	R_E, kΩ	R_C, kΩ	β_{dc}	I_C, mA	V_{CE}, V
12	1	1.2	100	3	5.4
12	1	1.2	400	3	5.4
12	1	1.8	100	3	3.6
12	1.5	1.2	100	2	6.6
18	1	1.2	100	3	11.4

 b Remains the same
 c Remains the same

 d Remains the same
 e Decreases
 f Decreases
 g Increases
16 Opened = 0
    Closed = 20 mA

17

V_{BB}, V	R_E, Ω	Diode Current, mA
0	330	0
4	330	10
4	1 k	3.3

18

Circuit Values, 10% decrease	Circuit Response			
	Base Bias		Emitter Bias	
	I_C, mA	V_{CE}, V	I_C, mA	V_{CE}, V
β_{dc}	D	U	N	N
V_{CC}	N	D	N	D
V_{BB}	D	U	D	U
R_C	N	U	N	U
R_B	U	D	XXXXX	XXXXXXX
R_E	XXXXX	XXXXXXX	U	D

19

Trouble	Voltmeter Readings			
	V_1, V	V_2, V	V_3, V	V_4, V
Circuit OK	30	5	4.3	15.8
R_C shorted	30	5	4.3	30
R_C opened	30	5	4.3	4.3
Base-emitter diode opened	30	5	0	30
Base-collector diode opened	30	5	4.3	30
All transistor terminals opened	30	5	0	30
All transistor terminals shorted	30	5	5	5
Base supply voltage off	30	0	0	30
Collector supply voltage off	0	5	4.3	0

Chapter 8

1a

R_E, kΩ	R_C, kΩ	β_{dc}	I_C, mA	V_{CE}, V
3.3	10	100	1.16	14.6
6.8	10	100	0.565	20.5
3.3	18	100	1.16	5.29
3.3	10	200	1.17	14.4

b Decreases
c Increases
d Remained the same
e Decreases
f Remains the same
g Remains the same
2 Circuit *a*
3a 11.7 V
 b 1 mA
 c 10 kΩ
4 Circuit *a*

5a

R_E, kΩ	R_C, kΩ	β_{dc}	I_C, mA	V_{CE}, V
5.6	3.3	100	3.44	9.38
10	3.3	100	1.93	14.3
5.6	4.7	100	3.44	4.57
5.6	3.3	200	3.44	9.38

b Decreases
c Increases
d Remains the same
e Decreased
f Remains the same
g Remains the same
6 Circuit *a*
7 Circuit *a*
8a 5.7 V
 b 10 mA
 c 1 kΩ

9a 1.66 mA
 b 1.66 mA
 c 5.22 V
 d 8.67 mW
10a 1.13 mA
 b −14.4 V
11a 1.88 mA
 b −11.6 V
12a 0.93 mA
 b −6.33 V
13a 9.3 mA
 b −9.54 V
14a 2.93 mA
 b 10.1 V
 c 29.6 mW
15a 1 mA
 b 58.5
 c 2.2 kΩ
16a 1.4 mA
 b −16 V
 c 22.4 mW
17a 8.33 mA
 b 114
18a 2.53 mA
 b 12.6 V
 c 31.9 mW
19a 3.03 mA
 b 107
 c 3.3 kΩ

20

Trouble	Voltmeter Readings			
	V_1, V	V_2, V	V_3, V	V_4, V
Circuit OK	30	3.91	3.21	18
R_1 opened	30	0	0	30
R_2 shorted	30	0	0	30
Base-emitter diode opened	30	3.91	0	30
All transistor terminals opened	30	3.91	0	30
All transistor terminals shorted	30	5.53	5.53	5.53
R_C shorted	30	3.91	3.21	30
Collector supply voltage off	0	0	0	0

Chapter 9

1a 0.723 Hz
 b 7.23 Hz
 c 0.707 V
 d 1 V
2a 0.318 Hz
 b 3.18 Hz
 c 0.354 V
 d 0.5 V
3 796 μF
4a

C μF	R_L kΩ	V_G V_{rms}	Critical Frequency
10	1	1	9.95 Hz
100	1	1	0.995 Hz
10	6.8	1	2.15 Hz
10	1	10	9.95 Hz

 b Decreases
 c Decreases
 d Remains the same
5a 31.8 Hz
 b 318 Hz
 c 7.07 V
 d 0 V
6a 3.98 Hz
 b 39.8 Hz
 c 2.12 V
 d 0 V
7 796 μF

8a

C μF	R_L kΩ	V_G V_{rms}	Critical Frequency
10	1	1	31.8 Hz
100	1	1	3.18 Hz
10	6.8	1	18.3 Hz
10	1	10	31.8 Hz

 b Decreases
 c Decreases
 d Remains the same

9a

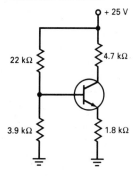

(a)

b

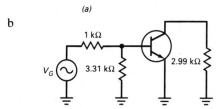

(b)

 c 1.7 mA
 d 14 V

10a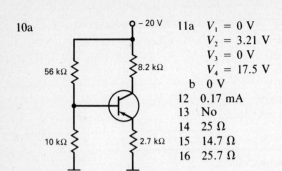

(a)

b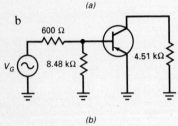

(b)

c 0.863 mA
d −10.6 V

11a $V_1 = 0$ V
     $V_2 = 3.21$ V
     $V_3 = 0$ V
     $V_4 = 17.5$ V
  b 0 V
12 0.17 mA
13 No
14 25 Ω
15 14.7 Ω
16 25.7 Ω

17a

R_E kΩ	R_C kΩ	R_L kΩ	I_E mA	V_{CE} V	r'_e Ω
1.8	4.7	10	2.47	13.9	10.1
3.3	4.7	10	1.35	19.2	18.5
1.8	8.2	10	2.47	5.3	10.1
1.8	4.7	22	2.47	13.9	10.1

  b Increases
  c Remains the same
  d Remains the same
18a 5 kΩ
  b 2.42 kΩ
19a $z_{in(base)} = 2.21$ kΩ
     $z_{in} = 1.33$ kΩ
  b $z_{in(base)} = 2.94$ kΩ
     $z_{in} = 1.56$ kΩ
  c It increases
  d It would increase
20 1.11 kΩ
21 2.1 kΩ
22 130

Chapter 10

1a 4 kΩ
 b 2 kΩ
2a 2.4 kΩ
 b 1.71 kΩ
3a 2 kΩ
 b 1.2 kΩ
4   1.28 kΩ
5   6.67 kΩ
6   150
7a 0.667 mV
 b 0.333 μA
 c 0.667 mV
 d 0.0333 mA
 e 100 mV
 f 150
8a 1.48 mV
 b 0.619 μA
 c 0.124 mA
 d 0.446 V
9a 1.5 mV
 b 0.5 V
 c 333
10a 2.16 mV
 b 1.69 μA
 c 0.214 mA
 d 685 mV
 e 317
11 250
12 144

13a

r'_e, Ω	R_L, kΩ	β	v_{in}, mV	A
15	3.3	200	2	129
30	3.3	200	2.25	64.7
15	10	200	2	213
15	3.3	400	2.25	129

  b Increases
  c Remains the same
  d Increases
  e Decreases
  f Increases
  g Remains the same
14 619 mV
15 344 mV
16a 15 kΩ
  b 6 kΩ
  c 8.57 mV
  d 66.7
  e 572 mV

17a

r_e, Ω	R_L, kΩ	β	v_{in}, mV	A
50	3.3	100	7.5	32.3
100	3.3	100	7.95	17.6
50	10	100	7.5	53.3
50	3.3	200	8	32.3

  b Increases
  c Remains the same
  d Increases
  e Decreases
  f Increases
  g Remains the same

18a 2.48 kΩ
  b 7.13 mV
  c 39.4
  d 281 mV
19   836 mV
20a 0.04 mV
  b 3.97 mV
  c 99.3
  d 31.5
  e 3137
  f 125 mV
21   210 mV
22a 0.863 mV
  b 27.3 mV
  c 31.6
  d 7.45
  e 235
  f 203 mV
23   225 mV

24

Trouble	Oscilloscope Readings				
	v_g, mV	v_b, mV	v_c, mV	v_e, mV	v_{out}, mV
Circuit OK	2	2	185	0	185
C_E opened	2	2	2.44	≈2	2.44
C_1 opened	2	0	0	0	0
C_2 opened	2	2	185	0	0
Voltage generator off	0	0	0	0	0
DC power supply off	2	2	0	0	0

Chapter 11

1a 16 V
 b 3 kΩ
 c 2 kΩ
 d 6 kΩ
2a 5.44 V
 b 5.51 V
 c Above
 d 10.9 V
 e Yes

3a 6.4 V
 b 50 mV
4   42 mV
    94.2 mV
5   89.1 mV
6a 6.03 mA
 b 109 mW
 c 5.5 mW
 d 24.8 mW
 e 5.05%

7a 4.36 mA
 b 52.3 mW
 c 1.02 mW
 d 0.925 mW
 e 1.95%
 f 1.77%
8a 3.97 mA
 b 60 mW
 c 2.67 mW
 d 2.15 mW

 e 4.45%
 f 3.58%
9a 400 mW
 b 200 mW
 c −2.67 mW/°C
10 Yes

Chapter 12

1a 24 kΩ
 b 4.8 kΩ
2a 42.4 kΩ
 b 8.09 kΩ
3a 14 kΩ
 b 2.47 kΩ
4   3.29 kΩ
5   6.67 kΩ
6   0.9
7a 82.8 mV
 b 3.45 μA
 c 0.69 mA
 d 69 mV
 e 0.833
 f 0.833
8a 124 mV
 b 2.92 μA
 c 0.585 mA
 d 117 mV
 e 0.943
 f 0.943
9   0.8
10a 0.884
 b 339 mV
11 0.942

12a

r'_e Ω	R_L Ω	β	z_{in} kΩ	A
5	22	200	2.56	0.81
10	22	200	2.78	0.683
5	100	200	3.93	0.946
5	22	400	3.39	0.81

 b Increases
 c Increases
 d Increases
 e Decreases
 f Increases
 g Remains the same
13 530 mV
14 324 mV
15a 0.964 V
 b 1.01 V
16a 2.06 V
 b 3.89 V
 c 1.19 V
17a 0.42 V
 b 0.648 V
 c 0.85 V

18a 628 Ω
 b 18.7 kΩ
 c 340
 d 0.973
 e 331
 f 662 mV
19 695 mV
20a 15.6 V
 b 1.82 V
 c 5.5 mV
21a 1.16 kΩ
 b 5.35 kΩ
 c 280
 d 0.979
 e 274
 f 736 mV
22a 14 V
 b 2.3 V
 c 15.6 mV
23a 9.79 mA
 b 97.9 μA
 c 7.4 V
 d 6.7 V
 e 5.1 Ω
24a 359 kΩ
 b 43.9 kΩ
 c 81.4 mV
 d 0.858
 e 69.8 mV
25 71.1 mV
26 51.8 mV

Chapter 13

1. 200 MΩ
2. 500 MΩ
3. a. 16 mA
   b. −4 V
   c. 4 V
   d. 40 V
   e. 250 Ω
4. a. 15 mA
   b. 0 mA
   c. 5 V
   d. 333 Ω

5. +5 V

6a.

V_{GS} V	I_D mA
0	20
−1	12.8
−2	7.2
−3	3.2
−4	0.8
−5	0

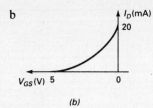

(b)

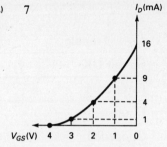

7

8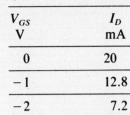

9. a. −4 V
   b. 500 Ω
   c. 8 V
   d. 4 mA
   e. 2 V
   f. 6.2 V
10. a. 2.46 mA
    b. 1.23 V
11. a. 4 mA
    b. −9.2 V
12. a.

V_{GS} V	V_{DD} V	R_D kΩ	I_D mA	V_{DS} V
−2	12	1	2.5	9.5
−1	12	1	5.63	6.37
−2	20	1	2.5	17.5
−2	12	2.2	2.5	6.5

   b. Increases
   c. Decreases
   d. Remains the same
   e. Increases
   f. Remains the same
   g. Decreases
13. I_{DSS} = 8 mA
    $V_{GS(off)}$ = −3 V

14. a. −3 V
    b. 500 Ω
    c. 6 V
    d. 3 mA
    e. 1.5 V
    f. 5.5 V
15. a. 1.64 mA
    b. 0.82 V
16. a. 16.3 mA
    b. 6 V
    c. 8.15 V
    d. 13.7 V
17. a.

V_{GS} V	V_{DD} V	R_D kΩ	I_D mA	V_{DS} V
−1	20	1	6.67	13.3
0	20	1	15	5
−1	30	1	6.67	23.3
−1	20	2.2	6.67	5.33

   b. Increases
   c. Decreases
   d. Remains the same
   e. Increases
   f. Remains the same
   g. Decreases

18a 7.11 mA
  b −9.34 V
19 0.25 mA/V²
20a 0.5 mA/V²
  b

V_{GS} V	I_D mA
1	0
2	0.5
3	2
4	4.5
5	8
6	12.5

c

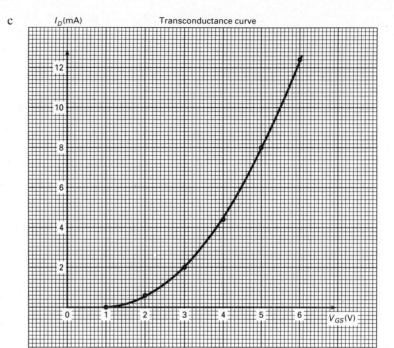

(c)

21a 1.56
  b 7.8 mA
  c 10.6 V
  d 1.56 V
22a +5 V
  b 4.5 mA
  c 8.25 V
  d 2.25 V
23a 2.88 mA
  b 1.44 V
24a 8 mA
  b −12 V

Chapter 14

1a 5.76 mA
  b 0 V
  c −1.9 V
  d 10.4 V
2a 4.26 mA
  b 11.4 V
  c 48.6 mW
3a 4.19 mA
  b −1.63 V
  c 1.4 V
  d 12 V
4 700 Ω
5a

R_S Ω	V_{DD} V	R_D kΩ	I_D mA	V_{DS} V
330	24	2.7	4.61	10
470	24	2.7	3.75	12.1
330	30	2.7	4.61	16
330	24	3.9	4.61	4.5

25a

V_{GS} V	V_{DD} V	R_D kΩ	I_D mA	V_{DS} V	V'_K V
7	16	1.5	4	10	2
8	16	1.5	6.25	6.63	3.13
7	24	1.5	4	18	2
7	16	2.7	4	5.2	2

  b Decreases
  c Increases
  d Remains the same
  e Increases
  f Remains the same
  g Decreases
6a 5.85 mA
  b −1.58 V
  c 1.46 V
  d 7.61 V
7a 3.87 mA
  b −1.51 V
  c 1.55 V
  d 13.4 V
8a 4.81 mA
  b −1.3 V

  b Increases
  c Decreases
  d Remains the same
  e Increases
  f Remains the same
  g Decreases
26 7500 MΩ
27 No

  c 0.963 V
  d 10.7 V
9a 10,000 μmho
  b 5000 μmho
  c 10,000 μmho
  d 7071 μmho
10a 6000 μmho
  b 3550 μmho
11a −2.5 V
  b 2.5 mA
12a 2.2 MΩ
  b 28.7 mV
  c 14.3
  d 410 mV
  e 115 μA
  f 28.7 mV

13a

g_m μmho	R_L kΩ	A	v_{in} mV
3000	5.6	6.9	47.6
4000	5.6	9.2	47.6
3000	12	8.82	47.6

  b Increases
  c Increases
  d Remains the same
  e Remains the same

14a 10 MΩ
  b 9.9 mV
  c 9.95
  d 98.5 mV
  e 35.1 μA
  f 9.9 mV
15 254 mV
16a 10 MΩ
  b 98 mV
  c 0.49
  d 48 mV
  e 150 μA
  f 50 mV

17a 2000 μmho
  b 45.5 mV
  c 0.563
  d 25.6 mV
  e 39.8 μA
  f 19.9 mV
18 54.3 mV
19a 1.96 mV
  b 100 mV
20a 19.8 mV
  b 12.9
  c 255 mV

21a 0.444 V
  b 20 V
22a 4.11 mA
  b 0 V
  c −1.43 V
  d 1.43 V
  e 8.9 V
  f 7.47
  g 5000 μmho
  h 3210 μmho

Chapter 15

1a 15 V
  b 19.3 mA
  c 3.7 V
2a $V_A = 0.7$ V
     $I_D = 28.6$ mA
  b $V_A = 0.7$ V
     $I_D = 6.6$ mA
  c $V_A = 2$ V
     $I_D = 0$ mA
3a 12 V
  b 0.7 V
  c 3 mA
  d 1.8 mA
  e 2.93 mA
  f 4 mA
4a 12 V
  b 0.7 V
  c 30 mA
  d 18 mA
  e 29.3 mA

  f 29.3 mA
5a 30V
  b 0.7 V
  c 5.7 V
6a 10 V
  b 0.7 V
  c 10.64 V
7a $I_G = 0$ A
     $I_L = 0$ A
     $V_{out} = 15$ V
  b $I_G = 4$ mA
     $I_L = 6.5$ mA
     $V_{out} = 0.7$ V
  c $I_G = 0$ A
     $I_L = 6.5$ mA
     $V_{out} = 0.7$ V
8a $I_G = 0$ A
     $I_L = 0$ A
     $V_{out} = 15$ V
  b $I_G = 0.851$ mA
     $I_L = 0$ A
     $V_{out} = 15$ V

  c $I_G = 0$ A
     $I_L = 0$ A
     $V_{out} = 15$ V
9a $I_G = 0$ A
     $I_L = 0$ A
     $V_{out} = 15$ V
  b $I_G = 4$ mA
     $I_L = 4.33$ mA
     $V_{out} = 0.7$ V
  c $I_G = 0$ A
     $I_L = 0$ A
     $V_{out} = 15$ V
10a 747 Ω
  b 1.93 kΩ
11 11.75 V
12a 12 V
  b 6.22 V
  c 10.6 mA
  d 4.76 mA
  e 19.5 mA

13a 0 A
  b 143 mA
  c 143 mA
  d Increased
14a 19.3 mA
  b 9.3 mA
  c 5 mA
15

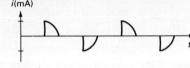

16a 0.6 V
  b 9 V
  c 9.7 V
  d Greatly reduced
17a 7.2 V
  b 3.5 V
18a 12.7 V
  b 1 V

Chapter 16

1 50 Hz to 25 kHz
2 1.41 V
3 2 Hz and 300 kHz
4a

C_{in} μF	z_{in} kΩ	R_G kΩ	f_c Hz
10	4	1	3.18
20	4	1	1.59
10	8	1	1.77
10	4	2	2.65

  b Decreases
  c Decreases
  d Decreases
5 14.5 μF
6a 90.4 Hz
  b 72.3 Hz
7a 41.1 Hz
  b 36.4 Hz
8a 21.2 Hz
  b 10.6 Hz

9 10.6 μF
10 10.2 Hz
11 140 Hz
12 63.7 Hz
13 318 μF
14 91.5 Hz
15 480 Hz
16 1.26 kHz
17 1.47 kHz
18a 4.78 MHz
  b 2.39 MHz
19 95.6 pF
20 8.87 MHz
21 1320 MHz
22a 1010 pF
  b 10.1 pF
23a 1510 pF
  b 10.1 pF
24 31.5 kHz
25 1.97 MHz
26 105 kHz
27 3.5 MHz
28 159 pF
29 50 Ω
30a 79.6 kHz
  b 637 kHz

31a 1384 pF
  b 24 pF
  c 215 kHz
  d 2.07 MHz
  e 215 kHz
32a 25 Ω
  b 63.7 pF
  c 403 pF
  d 18 pF
  e 561 kHz
  f 3.16 MHz
  g 560 kHz
33 285 kHz
34a 0.316 V
  b 0.707 V
  c 0.894 V
  d 0.894 V
  e 0.707 V
  f 0.316 V
35a 3 dB
  b 10 dB
  c 13 dB
  d 20 dB
  e 27 dB
  f 30 dB

36a 1000
  b 30 dB
37a 2
  b 4
  c 10
  d 100
  e 316
  f 10,000
38a 6 dB
  b 20 dB
  c 26 dB
  d 40 dB
  e 54 dB
  f 60 dB
39a 400
  b 52 dB
40a 1.41
  b 2
  c 3.16
  d 10
  e 17.8
  f 100
41a 5000
  b $A'_1 = 34$ dB
     $A'_2 = 40$ dB
  c 74 dB

42a 50 dB
b $A_1 = 10$
  $A_2 = 31.6$
c 316

43 A'_v(dB)

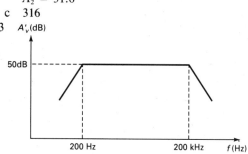

44 A'_v(dB)

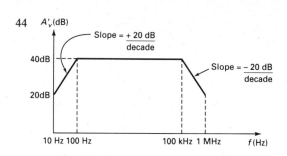

Chapter 17

1a 19.3 μA
b 9.65 μA
c 10.35 V
2a 20 μA
b 10 μA
c 10 V
3a 19.1 μA
b 9.55 μA
c 10.45 V

4a 193 μA
b 96.5 μA
c 10.35 V
5 0 V
6a 10 nA
b 70 nA
7 53 nA, 47 nA
8a 150
b 334 kΩ
c 1.5 V

9a 100
b 500 kΩ
c 1 V
10a 143
b 175 kΩ
c 2.86 V
11 2.5
12a 0.5
b 25 μV

13a 200
b 0.5
c 1200 mV
d 2 mV
e 1202 mV
f 400
g 52 dB
14a 173
b 0.5

c 1730 mV
d 1 mV
e 1731 mV
f 346
g 50.8 dB
15 286
16 2

Chapter 18

1a 1.59 kHz
b 159 kHz
2a 1.59 kHz
b 199 kHz
3 25 Hz
4 160,000
5 6 MHz
6a

V_P V	f kHz	S_S V/μs
10	20	1.26
20	20	2.51
10	40	2.51

b Increases
c Increases
7 2 mA

8 1 V/μs
9 10 pF
10 Yes
11 No
12 1.26 V/μs
13a

V_P V	S_R V/μs	f_{max} kHz
5	1	31.8
10	1	15.9
5	2	63.7

b Decreases
c Increases
14 318 kHz
15 3.18 V
16 4 V
17 10 μV

Chapter 19

1a 0.005
b 2500
c 1 V
d 2 μV
2a 4 V
b 10 μV
c 50 mV
d They are the same.
3a 3.9992 V
b 9.998 μV
c 49.99 mV
d They can be considered the same.

4a

R_2, kΩ	R_L, kΩ	A	v_{out} V	v_{error} μV
1	10	100,000	1.6	16
3	10	100,000	0.54	5.4
1	20	100,000	1.6	16
1	10	200,000	1.6	8

b Decreases
c Remains the same
d Remains the same
e Decreases
f Remains the same
g Decreases

5a 10 V(dc)
b 25 μV(dc)
6a 500,000
b 200
c 2 V
d 4 μV
7a 400,000
b 80
c 800 mV
d 2 μV
8a 2001 MΩ
b 0.05 Ω
9a 2501 MΩ
b 0.032 Ω

10a 10 V
  b 2 mA
  c 10 mA
11a 2 V
  b 20 μV
  c 0.02 Ω
  d 0.0008 Ω
12 5 V
13 1.2 kΩ
14 0.8 mA
15a 40 Hz
  b 20 kHz
16a 50 kHz
  b 250,000
17a 2 MHz
  b 10 kHz
  c 2 MHz
18a 316,000
  b 7.9 MHz
19a 158,000
  b 50.6 Hz
  c 200 kHz
  d 8 MHz
20 3.98 V

Chapter 20

1a 2.18 V
  b 44.8 kHz
  c 0.332 Hz
     1.59 Hz
     0.677 Hz
2a 1.01 V
  b 25 mV
3a 0.0099
  b 100
  c 5 V
  d 50 μV
  e 1.8 kΩ
  f 0.101 Ω
  g 50 kHz
4a 0.5 V
  b 100 Ω
  c 0.5 Ω
  d 10 kHz
5a

R_S kΩ	R_F kΩ	A	v_{out} V	v_{error} μV
1	50	100,000	5	50
2	50	100,000	2.5	25
1	100	100,000	10	100
1	50	200,000	5	25

  b Decreases
  c Increases
  d Remains the same
  e Decreases
  f Increases
  g Decreases
6 10 V(dc)
7a 0.5 V
  b 0.1 mA
  c 0.1 mA
8a 1 kΩ
  b 80 kΩ

9a 100
  b 5 V
  c 4.98 kHz
10a 31.8 Hz
  b 15.9 Hz
  c 6.37 Hz
  d 2 V
11 3 V(dc)
12 700 mV
13a 5.1 V(dc)
  b 51 mA(dc)
  c 0.225 mA(dc)
14a 5.1 V(dc)
  b 510 mA(dc)
  c 2.25 mA(dc)
15a

R kΩ	V_{in} V(dc)	R_L Ω	i_{out} mA(dc)
1	1	100	1
2	1	100	0.5
1	2	100	2
1	1	200	1

  b Decreases
  c Increases
  d Remains the same
16a

R kΩ	V_{in} V(dc)	R_L Ω	i_{out} mA(dc)
1	5	100	10
2	5	100	5
1	10	100	5
1	5	200	10

  b Decreases
  c Decreases
  d Remains the same